The Science Explorer Out and About

**Fantastic Science
Experiments Your
Family Can Do
Anywhere!**

Pat Murphy, Ellen Klages, Linda Shore,
and the Staff of the Exploratorium

Illustrated by Jason Gorski

An Owl Book   Henry Holt and Company   New York

Henry Holt and Company, Inc.
Publishers since 1866
115 West 18th Street
New York, New York 10011

Henry Holt® is a registered trademark of
Henry Holt and Company, Inc.

Library of Congress Cataloging-in-Publication Data
The science explorer out and about: fantastic science experiments
your family can do anywhere / Pat Murphy . . . [et al.].
        p.   cm.—(An Exploratorium science-at-home book)
    "An Owl book."
    Includes index.
    ISBN 0-8050-4537-6 (An Owl Book: pbk.)
        1. Science—Experiments. 2. Science—Experiments—
Juvenile literature.   3. Scientific recreations. 4. Science museums—
Educational aspects.   I. Murphy, Pat.   II. Series.
Q182.3.S335   1997   97-16847
507.8—dc20   CIP

Henry Holt books are available for special promotions and premiums.
For details contact: Director, Special Markets.

First Edition 1997

Printed in the United States of America
All first editions are printed on acid-free paper. ∞

10   9   8   7   6   5   4   3   2   1

Be careful! The experiments in this publication were designed with safety and success in mind. But even the simplest activity or most common materials can be harmful when mishandled or misused.

Exploratorium® is a registered trademark and service mark of The Exploratorium.

Air Crisps® is a registered trademark of Nabisco Brands Co.
X-Acto® is a registered trademark of Hunt Manufacturing Co.
Mrs. Stewart's® is a registered trademark of Luther Ford & Co.
Mini-MagLite® is a registered trademark of Mag Instruments.
Lava Lite® is a registered trademark of Lava-Simplex Internationale, Inc.
Play-Doh® is a registered trademark of Tonka Corporation.
Softsoap® is a registered trademark of Colgate-Palmolive Co.
Cheerios® is a registered trademark of General Mills, Inc.
Post-it® is a registered trademark of 3M.
Ivory Soap® is a registered trademark of Procter & Gamble.

Photo Credits
All photos are by Exploratorium staff photographer Amy Snyder unless otherwise indicated here. Pages 1, 36, 81 (right): Susan Schwartzenberg; Page 21: Hal Beilan; Page 65: Esther Kutnick; Page 99: Courtesy Mary K. Miller.

This book is dedicated
to the spirit of curiosity
that brings out
the natural scientist
in everyone.

Contents

Finding the Science All Around You

An Introduction for Parents

Children are natural scientists. Given half a chance, a kid can turn a leisurely walk into a science safari—searching for bugs, looking at birds, finding every fern and flower along the way. Take a child to the beach and she's just as likely to spend hours sifting through the sand, examining the tiny colored rocks, as she is to be playing in the surf. Even at night, when you think a kid is finally falling quietly asleep, he may be singing into his pillow to see what kinds of weird sounds he can make, or making shadow pictures on the wall.

Children are experts at figuring out how the world works—examining their environment, testing their ideas, asking endless questions, and experimenting to try to figure out the answers. With this book, you can encourage that natural enthusiasm, helping your child develop and maintain the creativity and curiosity that makes the world such an exciting place to explore.

The Exploratorium's Science-at-Home Team has created dozens of intriguing activities for this book. With the help of hundreds of families worldwide, each experiment has

The Science-at-Home Team—Pat, Ellen, and Linda.

been tested to make sure it works—and to make sure it's fun! We've written simple, clear instructions to ensure that all your experiments will work as expected. We've included straightforward explanations to help you answer questions about what's going on. And we've made sure that all the ingredients are easy to find—if you don't already have them at home, you'll be able to find them at the grocery store.

This book has experiments that kids can do on their own, and experiments that families can do together. Some activities are perfect for beautiful, sunny days; others you can do on those dreary, rainy days, when everyone's tired of doing the same old things. There are activities that just take a few minutes, and others that can provide a whole afternoon of excitement and discovery. At the end of each chapter are suggestions for a family field trip—the perfect way for Science Explorers to get out and about.

In short, there's something here for everyone, and for any time or place. Have fun exploring!

Jason (our illustrator) says "Hi!"

How to Use This Book

How to Be a Science Explorer

Like *The Science Explorer,* the first book in this series, there's no right or wrong way to use *The Science Explorer Out and About.* You can do the experiments section by section, or skip around in the book, playing with mirrors one day, and building gumdrop towers the next.

But if you want some help choosing just the right activity for just the right time, we've done our best to make things easy for you. There are things kids can do by themselves, things kids and grown-ups can do together, and (when nobody's looking) things grown-ups can do all by themselves!

Yes, You're Really Doing Science!

You don't need any special science background to do the activities in this book. All you need is a bit of curiosity and a willingness to experiment and try new things. After all, you never know what you might discover!

While you're experimenting, it may look (and feel) like you're just fooling around. You may be trying a little of this, doing a little of that. This approach to the world is at the heart of the scientific process. Many scientific discoveries have come about because someone was "just fooling around."

And don't worry if an activity doesn't do exactly what you thought it was going to do. You can learn just as much from the things that don't work as you can from the things that do. In fact, the history of science is filled with accidental discoveries—from Post-it notes to penicillin.

The most valuable lessons to be learned here are not to be found in the facts and figures that explain what's going on, but in recognizing that examining the world around you—observing, experimenting, asking questions, testing ideas—can provide even more important tools for intellectual growth. Encouraging a natural instinct to investigate and explore can make a lifetime of difference in the way a child looks at the world.

Kids Can Do It!

Activities labeled Kids Can Do It were designed for kids who can read and want to experiment on their own. They may need an adult to help get them going, but once they figure things out, don't be surprised if they're happy to work by themselves.

Exploring Together

Activities labeled Exploring Together were created especially for grown-ups and kids playing side by side. These activities encourage the whole family's participation. Working together, you can complete projects that require more manual dexterity or involve tools that you may not want kids using by themselves.

Family Field Trips

We've added a new section to *The Science Explorer Out and About* that didn't appear in *The Science Explorer*. At the end of each chapter, Family Field Trips take you and your family out into your neighborhood, suggesting new ways to explore the topics in each chapter. Our test families reported that getting out into their neighborhoods changed the way they looked at the world around them—and gave them ideas for family activities that they might not have considered before.

Finding What You'll Need

The materials you'll need are clearly listed at the beginning of each activity. We've also listed materials used throughout the book on pages xiv–xv. We did our best to find the simplest, safest, most available ingredients, but don't be afraid of experimenting on your own and trying out different materials. If you don't have colored paper, for instance, maybe you can write with colored crayons (see Silly Sentences, page 92). If something calls for sugar cubes (see Edible Architecture, page 30), but you've got a big box of croutons in the house, give them a try! You never know what wonderful new creations you may be able to make.

How Long Will It Take?

Some activities in this book are quick to do, great for when you just have a few minutes. Others take a bit longer, and may be just the thing to perk up that dreary rainy day. We've marked each experiment with a clock that indicates about how much time it will take:

This means 15 minutes or less

This means 15 minutes to 1 hour

This means 1 hour or more

This is based on our best estimate of the shortest time needed to do the experiment. If an activity intrigues you, you may find yourself spending more time—or going back to do it again and again.

What's Going On?

This book includes explanations that grown-ups can use to answer questions that may come up while the kids are experimenting. These explanations, found in boxes called "What's Going On?", are written for adults (or older kids) who have no scientific background.

Explanations Are Optional

We suggest that grown-ups read any explanations they might find interesting, and share them with their kids if they ask. Some children may want explanations, but many others are just interested in playing around and experimenting. Rather than trying to explain, we advise grown-ups to encourage a child's natural tendency to figure out what will happen when he or she tries something new and different.

A Note on Grammar

The Science-at-Home Team is dedicated to encouraging both girls and boys to experiment and to explore their world. We were concerned that subtle cultural biases tend to discourage girls from pursuing an interest in science. With that in mind, we found ourselves in a dilemma. We did not want to use only "he" or "she" to refer to an unspecified friend or helper. But we also didn't want to invent a new pronoun or resort to grammatical errors like "Have your friend put their penny in the salt water." Our solution was to use "she" sometimes and "he" sometimes, so that some of your helpers are girls and some are boys.

Discovering on Your Own

The instructions in this book can help you start experimenting. But your explorations don't have to stop there. You can approach our instructions the same way a creative cook approaches a recipe. Change a little something and see what happens!

One of the things we found in the course of creating this book is that you never know where a great idea will come from. These activities have been checked by hundreds of families around the world, and with each round of testing we were always delighted to find that people discovered things we hadn't expected or anticipated. Those discoveries helped shape this book.

Let Us Know What You Think

The process of discovery is always full of surprises. Did you find something you'd like to share with us? Did a particular project make a difference to someone you know? Did you come up with a new way to do one of the activities in the book? Did you create an activity of your own? We're excited about receiving your comments and suggestions and would love to hear from you.

You can contact us at:
The Science Explorer Out and About
Henry Holt and Company
115 West 18th Street
New York, NY 10011

Or you can send us e-mail at:
home_science@exploratorium.edu

Let us know what you think!

Where Should We Start?

Ready, Set, Explore!

This book has experiments and activities that are fun in a variety of situations. Use these lists to help you choose the perfect activity for the time you have, the materials you have, or the number of people you want to entertain.

We've Only Got a Few Minutes— What Can We Do?

If you don't have much time, you may want to try these quick experiments:

The Whole Gang Is Coming— What Can We Do?

Maybe you have an entire birthday party or scout troop to entertain. These experiments work well with groups:

It's Beautiful Today— What Can We Do Outside?

Maybe it's a sunny day and you want to be outside. Here are some experiments that are best done outdoors:

It's One of Those Yucky Rainy Days—What Can We Do Indoors?

If it's nasty out and you're trapped in the house, here are some great indoor experiments to try:

We Love Art!—What Can We Do That's Artistic?

Maybe you've got kids who love art but aren't so sure about science. Here are some experiments that might spark their curiosity:

What Will We Need?

There's nothing worse than being all ready to do an experiment—and then realizing you don't have everything you need. To help you plan ahead, we've listed materials you may have around the house, materials you may need to buy at the store, and special items you may need to help kids find and use. If kids (or grown-ups) want to be ready for anything, try putting together your own Instant Science Lab Box. You'll find instructions on page xvi.

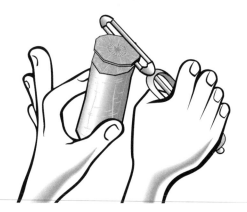

What Can We Find in the Kitchen?

Some experiments use food and equipment from the kitchen. If you don't already have these items around the house, they should be easily available.

banana
bowls
butter knife (not a sharp knife)
carrot
egg (raw)
food coloring
glass measuring cup
measuring spoons

vegetable oil
orange
paper towels
vegetable peeler
can of soda (full)
can of diet soda (full)
box of confectioners' sugar
box of sugar cubes
clean dish towel
bottle of white vinegar
wax paper or plastic wrap

What Will We Need from the Store?

For a few experiments, you'll need things that you may not usually keep around the house. You may need to buy these things from the store:

ammonia
rubbing alcohol (isopropyl)
a bottle of Mrs. Stewart's Bluing
a package of Nabisco Air Crisps crackers
1/2 pint of whipping cream
a bag of gumdrops
liquid soap with glycol stearate in it
    (like Colgate-Palmolive's Softsoap)
Epsom salts
charcoal briquettes

What Special Materials Will Kids Need Help Finding?

For some experiments, you may need to help kids find or use special materials, such as:

ammonia
a bottle of Mrs. Stewart's Bluing
a handful of small beads, sequins, or glitter
charcoal briquettes
a 3.5" computer disk
Epsom salts
300–400 pennies
penlights or Mini MagLites
rubbing alcohol (isopropyl)
screwdriver
utility knife or X-Acto knife
white sheet or tablecloth
wineglass

Which Experiments Require Advanced Planning?

If you want to plan ahead, we suggest making photocopies for two activities:

page 11 (Secret Shades)
page 69 (Making a Sun Clock)

How Can We Make an Instant Science Lab Box?

When the Science-at-Home Team first started experimenting, we spent lots of time hunting around for mirrors and scissors and toothpicks and clay. We finally got smart and put all the basic ingredients for these experiments into a big cardboard box: our Instant Science Lab Box. You can make one, too! Just put these items in a big box and you'll have on hand a lot of what you'll need for the experiments in this book. (Note: If something has this symbol • next to it, it means you'll have to plan ahead to save up some of this stuff, especially if your family does a lot of recycling.)

- • aluminum pie pan
- • brown paper grocery bag
- • white plastic grocery bag
- colored chalk

modeling clay
pad of construction paper
box of crayons
- • 2 baby food jars
- • a glass jar with a lid
- • a few different-sized envelopes
package of file cards (with lines)
- • sheets of thin cardboard or posterboard
markers (black and colored ones)
2 small mirrors (square ones work best)
a couple of steel nails and screws
- • a newspaper
a nickel
white paper (notebook size)
a few paper clips
pen
pencil
- • clear plastic containers with lids
- • a handful of rubber bands
a ruler
scissors
- • shoebox
a sponge
string
tape (clear plastic tape and masking tape)
a box of round toothpicks

Have Fun!

It's All Done with Mirrors

Mirrors can be strange and tricky things. At the Exploratorium, there's a mirror that will even make it look like you can fly.

At *Anti-Gravity Mirror,* a visitor stands on a block at the far edge of a long mirror. Her left leg is behind the mirror and her right leg is in front of the mirror. The edge of the mirror is at the center of her body. Her friends watch the front of the mirror.

When she lifts her right leg, it looks like she's flying! Both of her legs appear to be completely off the ground, and her whole body looks like it's suspended in midair.

The trick is that her friends—who think they're seeing her whole body—are really only seeing her right side and its reflection in the mirror. Her two right halves look like they make a whole body. That's because people's bodies are *symmetrical,* which means both sides look the same. On each half of your body you have an arm, a leg, and an eye.

The two halves aren't exactly the same, but they look enough alike to fool your brain into seeing a whole body in the mirror.

In this chapter, you can use mirrors to experiment with the symmetry of shapes and letters. Make a periscope so you can see around corners, or see what happens when a mirror turns the alphabet upside-down. Learn where to look for strange reflections in your neighborhood. Find out how a mirror can fool your brain, making it hard to tell if something's going left or right or forward or backward. You may even forget how to write your own name!

Hat Trick

Can you write your name? Can you draw a happy face? Want to bet?

What Do I Need?

- A small mirror
- Paper
- Pencil, pen, or marker
- A table
- A baseball cap (if you've got one)
- Masking tape or other strong tape
- A maze from an activity book (if you've got one)

What Do I Do?

If You Have a Baseball Cap
Tape the mirror to the underside of the brim of the cap. Use enough tape to hold the mirror on tight. Put on the cap. Look up into the mirror. Adjust the cap so that when you're looking up into the mirror you can see the paper on the table.

If You Don't Have a Baseball Cap
If you're right-handed, use your left hand to hold the edge of the mirror against your eyebrows. If you're left-handed, use your right hand to hold the mirror. It's easier to hold the mirror for a long time if you rest your elbow on the table. Look up into the mirror to see the paper on the table.

1 Put a piece of paper on the table. Look up into the mirror. Watch your hand in the mirror and try to print your name on the piece of paper. Don't look down at your hand or the paper. Just look up into the mirror. Print your name so you can read the letters in the mirror.

Wow! I Didn't Know That!

Italian painter and inventor Leonardo da Vinci kept a journal of his ideas and inventions. To keep his thoughts private, he wrote backwards, like this:

I'm going to paint Mona Lisa tomorrow.

When da Vinci held his journal up to a mirror, he could read the reflection of his backward words. Try it and see for yourself!

2 If you've written your name a couple of times, try writing the word *Exploratorium.* Or try writing a whole sentence.

3 Try to draw something simple, like a happy face or a house. Even that may be really hard to do! Most of the time when you write or draw, your eyes and hands work together. But when you look in the mirror, what your eyes see and what your hands do aren't the same.

Home Scientists in the Judd/Clear family loved this activity. They said it felt really weird to see the floor on the ceiling!

4 For a real challenge, put a maze down on the table. Then look up into the mirror and try to draw a line from the start to the finish. You won't believe what your hand will try to do!

What's Going On?

Why are drawing and writing so difficult when I'm looking in the mirror?

In this experiment, you set up a situation where the information you get from your eyes doesn't match the messages you get from your hand. When you look up into the mirror, what you see on the paper is upside-down. When you try to move your hand up, the mirror shows it moving down. You see your hand moving in one direction—but you feel yourself moving it in another.

The conflict between your vision and your sense of where your hand is can be almost paralyzing. It's a little like trying to learn to write for the first time. Learning to write is hard at first, because your brain has to learn to tell your hand exactly what to do. With practice, you learned to coordinate your brain and hand. Now, looking in the mirror, you are trying to write in a mirror world, where everything is reversed. The usual signals sent by your brain send your hand in the wrong direction. With enough practice, you can learn to write in a mirror world—but it isn't easy!

Flipping Shapes

Use a mirror to turn a fish into a bow tie!

What Do I Need?

- A small mirror (a square mirror works best)
- The pictures in this activity
- Pencil, pen, or marker
- Paper

What Do I Do?

1 Put your mirror on the faint line that goes up and down through the middle of the face, between the eyes. Do you see another face in the mirror? Try the same thing with the fish and the flower. Do you see the same fish and flower, or new pictures?

2 Now put your mirror on the line that goes across the fish, through the middle of its tail. What do you see in the mirror? Try putting your mirror on the same line through the face and the flower. What happens?

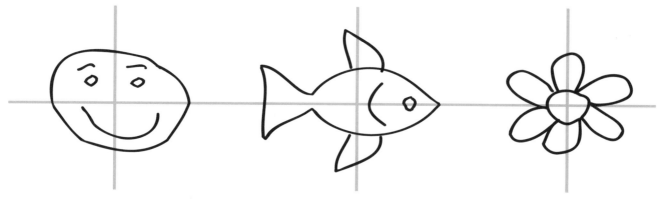

3 Here are 3 more pictures to play with. Can you turn the mirror so that the hot dog turns into a heart? The heart turns into a butterfly? A diamond? The fish turns into a bow tie? A triangle?

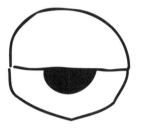

4 Can you draw other pictures that will change when you move the mirror? You might want to start by just making some squiggles and moving the mirror around to see what new shapes you can find. What new shapes can you find in these pictures?

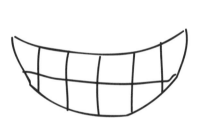

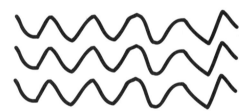

What's Going On?

About symmetry and mirrors

Reflecting pictures in a mirror is one way to experiment with a concept that mathematicians call *symmetry*. If you can draw a line through a shape so that everything that's on one side of the line has a twin on the other side, then that shape is *symmetrical*. If you put a mirror on that line, you get an image that looks like the original shape. This kind of symmetry is called *mirror symmetry*. It's also sometimes called *line symmetry* or *bilateral symmetry*.

What else can I do with symmetry?

Experimenting with symmetrical shapes can get you started on an exploration of symmetry in the natural world. Take a look at the plants and animals around you, and you'll find many examples of natural symmetry.

Take your own body, for example. You've got a right hand and a left hand, a right eye and a left eye, and so on. You, like many animals, are more or less symmetrical from side to side.

You're not perfectly symmetrical, however. If you're like most people, your heart is a little bit to the left of a central line. (In a few people, the heart is a little to the right of a central line, and all their other organs are reversed as well, a condition known as *inverse situs*.) Though your face looks symmetrical, it isn't really. If you place a mirror down the center of a picture of someone's face, the resulting image (made up of half the face and its reflection) will look subtly wrong.

Alphabet Flip

In a mirror, some letters of the alphabet look very strange, but some stay the same.

What Do I Need?

- A small mirror (a square mirror works best)
- The letters and words in this activity
- A wall mirror or bathroom mirror for step 5
- Pen or pencil
- Paper

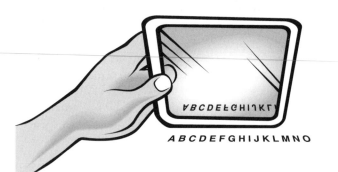

What Do I Do?

1 Put your small mirror on the dotted line above the alphabet at the bottom of the page. Look in the mirror. The letters in the mirror are all upside-down. Most of them look really strange that way. But some of them look okay. There are 9 letters that look the same right-side-up as they do upside-down. Can you find them all? Write those letters on a piece of paper. When you're done, you can check the answers at the end of the "What's Going On?" section on page 7.

ABCDEFGHIJKLMNOPQRSTUVWXYZ

2 Put a mirror on the dotted line above the words below. Some of the words look okay in the mirror, and some of them don't. The words that look okay are the ones that use letters from the list you made.

3 What other words can you make out of the letters on your list? Try writing them on a piece of paper and putting a mirror above them. Can you read the words in the mirror?

BIKE BOOK BROKE CHOICE HOOD OBOE ELBOW ECHO

4 You can use words that don't change to make whole sentences that don't change when you look at them in a mirror. Put the mirror on the dotted line above the sentence below. If you want, try to make up some sentences of your own.

5 Hold up this page in front of your bathroom mirror. Look in the mirror and try to read the sentence about Heidi. Now the sentence looks like nonsense. But if you turn the page upside-down, then look in the mirror again, you can read the sentence just fine!

HEIDI HICKOK DECIDED: HIDE ICEBOX COOKIE

What's Going On?

I can read some of the letters in the mirror. Why can't I read all of them?

If you put a mirror along the line in the middle of this H, you get an image that looks like the original H. That's because the letter H is *horizontally symmetrical*—the top half and the bottom half are mirror images of each other.

When you put a mirror above the alphabet, the image in the mirror shows you letters that are upside-down and backward. A letter that is horizontally symmetrical will come out of this transformation looking just like the original letter.

There are nine letters that are horizontally symmetrical. You can find them by putting your mirror above the alphabet printed in this activity and finding the letters that still look okay. (Answers are at right.)

What else can I do with these letters?

If you put a mirror on the dotted line below, you can read the sentence you see in the mirror.

HE DECODED—I DECIDED

Now suppose you hold the same sentence up to the bathroom mirror. You can't read it because the letters (and the sentence) will read backward—but not upside-down. To read the sentence in the bathroom mirror, you need to turn the book upside-down—so that the letters (and the sentence) are both upside-down and backward.

Answers

Hold this page up to the mirror to read the answers.

The nine letters that look the same right-side-up as they do upside-down are B, C, D, E, H, I, K, O, and X.

Flip-Flop

Use a mirror to change TOM into MOM.

What Do I Need?

- A small mirror (a square mirror works best)
- The words and letters printed in this activity

What Do I Do?

1 Put your mirror on the dotted line that runs through the center of the letter O. The mirror should be facing the letter T. What word do you see in the mirror?

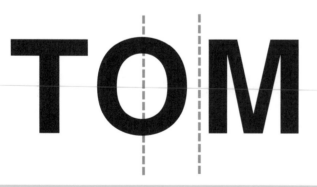

The Bulaevsky family sent us this great palindrome, which uses only vertically symmetrical letters:

AIM A TOYOTA TATAMI MAT AT A TOYOTA, MIA

2 Now put the mirror on the dotted line next to the O. The mirror should still be facing the letter T. What new word do you see in the mirror?

3 Now turn the mirror around so that it's facing the letter M. Put it on the dotted line in the center of the O. What word do you see in the mirror?

All three of the words you saw in the mirror are *palindromes.* A palindrome is a word that reads the same going backward as it does going forward. Here are some other examples of palindromes:

MOM PEEP OTTO KAYAK ROTATOR

If you hold the mirror to the left or right of the word MOM or the name OTTO, they look the same in the mirror as they do on paper. But PEEP and ROTATOR and most other palindromes won't do that because you can't read their letters when they're flipped in the mirror.

4 Some lower-case letters do a very strange thing when they're flipped in the mirror— they change into other letters! Look at this bunch of letters. They don't spell a real word, do they?

biupil

Now hold the mirror to the right side or left side of the bunch. What word do you see in the mirror? Wow! This is a great trick to show your friends.

5 Here are two other bunches of letters that will change into real words in the mirror. Can you think of any others?

wolliq
tliup

6 Use your mirror to look at the alphabet below. Who knows what letter tricks you might be able to find!

abcdefghijklmnopqrstuvwxyz

What's Going On?

What's so special about the letters in TOM?

The capital letters T, O, and M are all *vertically symmetrical*. They're symmetrical across a line drawn up-and-down through their middle. That means you can put a mirror on this line and the image you get will look just like the original.

There are eleven capital letters (including T, O, and M) that are symmetrical across a vertical line. Hold the alphabet on page 6 up to the mirror and see if you can find them. (If you want to check your answer, we list the letters at the end of this section.)

How can I use these letters to write sentences that look okay when I reflect them?

You can't just write words using these letters, reflect them, and read the mirror image. In the mirror image, the letters look fine—but their order is backward. To make words that read in the mirror, you have to reverse the order of the letters, like this:

.TOOH OT TOH OOT M'I

If you're very clever, you could try to write a sentence that's a palindrome (where the order of the letters is the same going backward as it is going forward) that uses only vertically symmetrical letters.

Answers

Hold this page up to the mirror to read the answers.

The eleven letters that are symmetrical across a vertical line are A, H, I, M, O, T, U, V, W, X, and Y. The four letters that are both horizontally and vertically symmetrical are H, I, O, and X.

Secret Shades

Make your own mirrored sunglasses with the liner from a box of crackers.

What Do I Need?

- The silvery inner bag from a box of any kind of Ritz Air Crisps or Cheese Nips Air Crisps crackers (these are the brands we tried; you may find other products with silvery inner bags that will work)
- Scissors
- Photocopies of the sunglass frames on page 11 (If you can't make photocopies of the sunglass frames, you can still make Secret Shades by using tracing paper. See step 5 for instructions.
- Thin cardboard or posterboard
- Glue
- Clear tape
- Two rubber bands
- Empty bowl, bag, or container for the crackers
- Pen

What Do I Do?

1 Open the bag inside the box of crackers. Pour the crackers out of the bag. (Check with a grown-up. You might be able to snack on the crackers while you experiment!)

2 Cut the bag open and flatten it out. Wash off the grease and salt inside with soap and water.

3 The bag is a big piece of plastic that's coated with a thin layer of metal. The metal is what makes it shiny, like a mirror. If you look at the bag, you can see a blurry reflection of your face.

4 Hold the bag up to your face and look through it at something light or bright—like a window. The metal coating on the plastic reflects a lot of light, but it lets some through, too. You can see right through the bag!

3 1/2"

Don't cut this out of your book!

FOLD

4"

3 1/2"

Don't cut this out of your book!

FOLD

This Is Important!
Make photocopies of these glasses before you cut them out. Or, if you want, you can use the measurements to design your own.

5 To turn the plastic bag into a pair of sunglasses, you'll need to make frames. Glue a photocopy of the sunglass frames onto cardboard. (If you don't have a photocopy of the frames on this page, you can still make Secret Shades! Just trace the sunglasses onto tracing paper and then glue the tracing paper pattern onto your cardboard.)

6 Now, with your scissors, carefully cut out your sunglasses. Don't forget to cut out the eyeholes, too! (The easiest way to do the eyeholes is to poke a hole in the middle of each one with a pen, and then use the scissors to cut out to the edges.) If you need help, just ask a grown-up.

7 Cut out a piece of the shiny plastic from the cracker bag. Make sure it's big enough to cover both eyeholes. Tape the plastic to the cardboard frames.

8 Ask a grown-up to poke a hole at the end of each long piece of cardboard. Bend the long pieces so they'll fit behind your ears.

9 Cut both rubber bands so you have two rubber strings. Tie one end of each string through the holes you poked, then tie the strings together.

10 Go outside and put the sunglasses on. The rubber bands will make them fit tightly around your head so you can see out, but other people won't be able to see your eyes!

Tips for Home Scientists

Making Secret Shades might be a good activity for a party, since one cracker bag will make 10 to 12 pairs of sunglasses.

What's Going On?

Have you ever noticed that windows sometimes act like mirrors?

When you're standing in a brightly lit room and looking into a window onto a dark landscape, you'll see your own reflection, rather than the world outside. When light shines on a window, most of the light passes through the glass. About 4 percent of the light reflects from the front surface—where the air meets the glass—and another 4 percent reflects from the back surface—where the glass meets the air again. All in all, about 8 percent of the light that shines on a window reflects. The other 92 percent passes through the glass.

When you're in a brightly lit room at night and you look at the window, 8 percent of the light from the room reflects to your eyes. That 8 percent is much brighter than the 92 percent of the light that comes in from the dimly lit world outside. So you see the reflected light, not the transmitted light.

When you're inside on a sunny day, the opposite is true: the light coming in from the outside is much brighter than the reflected light. So you see only the outside world. The reflected light is still there, but the transmitted light washes it out so you can't see your reflection.

The best time to watch windows turn into mirrors is at twilight. If you're lucky, you might find a window where the light is balanced so that you can see through the window and see things that are reflected in the window at the same time.

What does that have to do with my Secret Shades?

The cracker bag is made of clear plastic coated with a thin layer of metal. This plastic-and-metal combination reflects a lot of the light that's shining on it, but some of the light travels through.

The cardboard frames of your Secret Shades block light and keep it dark near your eyes. When you look at something bright, the transmitted light from the outside washes out the reflected light, and you see the outside world.

There are special two-way mirrors that work like your Secret Shades. They are made of half-silvered glass, glass that's coated with a layer of silver just thick enough to reflect half the light shining on the glass. The rest of the light travels through. If the light is bright on one side of the glass and dim on the other, the half-silvered glass acts like a mirror on the bright side and like a window on the dim side.

I See You!

Use a mirror to see around corners or turn a friend upside-down!

What Do I Need?

- Two small mirrors
- A friend
- A chair you can hide behind
- Two rooms connected by a door that you can shut

What Do I Do?

1 Ask your friend to stand behind you or beside you. Hold the mirror and turn it until you can see your friend's face in it. Ask your friend to look at the mirror and tell you what she sees. No matter where you stand, if you can see her eye, then she can see your eye!

2 Go into the other room. (Ask your friend to stay where she is.) Shut the door almost all the way—just leave it open 4 or 5 inches. Hold the mirror in your hand and stick your arm out through the opening.

Move your hand until you can see your friend in the mirror. If she looks at the door, all she can see is your arm. But if she looks into the mirror, she can see your eye again.

3 Hide behind a chair. Hold the mirror a little above your head. Look up into the mirror and move it around until you can see your friend in it. Does she look right-side-up or upside-down?

If she looks at the mirror, can she see your eye? Ask her if your eye looks right-side-up or upside-down.

4 Get a second mirror. Hide behind the chair again. Hold one mirror over your head, and hold the second mirror down near your lap. This part's a little tricky. Look at the mirror in your lap. At the same time, move the mirror over your head until you can see your friend.

Does she look right-side-up or upside-down? Can she see your eye?

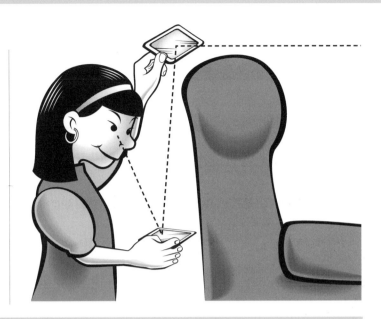

What's Going On?

Why can I see my face in a mirror?

To understand how mirrors work, you need to know about how light lets you see the world. You see these words because light is bouncing off this page and getting into your eyes. With that light, your eyes make an image of the page with the words on it.

You can see something only if light is bouncing off that thing and getting into your eyes. If you turn this book so it's facing away from you, you can't see the words any more. Light is still bouncing off the page, but it's bouncing away from you, and isn't getting into your eyes.

Now suppose you held the book up to a mirror. Try it and see.

You can see these words in the mirror even though the page is facing away from you. Light is bouncing off the page, just as it did before. The light that's bouncing off the page is shining onto the mirror. And the mirror is reflecting that light back into your eyes. That light gets into your eye, and you see the page.

When we play "I See You," why does my friend see my eye in the mirror?

Whenever you can see someone in a mirror, that person can see you in the mirror, too. You see the other person when light bounces off them, reflects from the mirror, and gets into your eyes. Unless you are standing in the dark, light is also bouncing off you. That light reflects from the mirror and gets into your friend's eyes—and your friend also sees you.

Light always reflects away from a mirror at the same angle that it hits the mirror. It's a little like a ball bouncing off a tennis court. If a ball comes in low to the ground, it bounces away low to the ground. If a ball comes in at a steep angle, it bounces high. The ball bounces away at the same angle that it hit the court. Or, as a physicist might say, the angle of incidence equals the angle of reflection.

Physicists call this the *law of reflection*. Because of the law of reflection, if you're standing at the correct angle to see your friend in the mirror, then your friend is standing at the correct angle to see you in the mirror, too.

A mirror is very smooth. When light from an object hits a mirror at an angle, it all bounces away at the same angle.

By the way, you just made a periscope.

A periscope is two mirrors that have been cleverly arranged. One mirror is angled so that it reflects an image of the world out there. The other mirror is angled so that it reflects the image from the first mirror into your eyes.

In step 4 of this activity, you made a periscope by holding the mirrors at the proper angle. Most periscopes have a tube that holds the mirrors. You can find out how to make your own periscope on page 16.

Up Periscope!

Build a mirrored tube that lets you see around corners and over walls.

What Do I Need?

- Two 1-quart milk cartons
- Two small pocket mirrors (flat, square ones work best)
- Utility knife or X-Acto knife
- Ruler
- Pencil or pen
- Masking tape

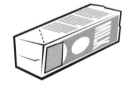

DANGER!
An X-Acto knife is very, very sharp. Have a grown-up do all the cutting in this activity.

What Do I Do?

1 Use the knife to cut around the top of each milk carton, removing the peaked "roof."

2 Cut a hole at the bottom of the front of one milk carton. Leave about 1/4 inch of carton on each side of the hole.

3 Put the carton on its side and turn it so the hole you just cut is facing to your right. On the side that's facing up, measure 2¾ inches up the left edge of the carton, and use the pencil to make a mark there. Now, use your ruler to draw a diagonal line from the bottom right corner to the mark you made.

4 Starting at the bottom right corner, cut on that line. Don't cut all the way to the left edge of the carton—just make the cut as long as one side of your mirror. If your mirror is thick, widen the cut to fit.

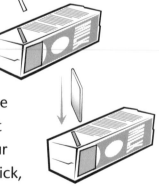

5 Slide the mirror through the slot so the reflecting side faces the hole in the front of the carton. Tape the mirror loosely in place.

Wow! I Didn't Know That!

Periscope comes from two Greek words, *peri,* meaning "around," and *scopus,* "to look." A periscope lets you look around walls, corners, or other obstacles. Submarines have periscopes so the sailors inside can see what's on the surface of the water, even if the ship itself is below the waves.

6 Hold the carton up to your eye and look through the hole that you cut. You should see your ceiling through the top of the carton. If what you see looks tilted, adjust the mirror and tape it again.

7 Repeat steps 2 through 6 with the second milk carton.

8 Stand one carton up on a table, with the hole facing you. Place the other carton upside-down, with the mirror on the top and the hole facing away from you.

9 Use your hand to pinch the open end of the upside-down carton just enough for it to slide into the other carton. Tape the two cartons together.

10 Now you have a periscope! If you look through the bottom hole, you can see over fences that are taller than you. If you look through the top hole, you can see under tables. If you hold it sideways, you can see around corners.

What's Going On?

What kinds of mirrors can I use to make a periscope?

You need two small mirrors, but they don't have to be identical. If you have a rectangular mirror, or one with a handle, it's okay if part of it sticks out the side of the carton. If your mirror is round, like the mirror in a make-up compact, you may want to tape or glue it to a square of cardboard before inserting it into the slot in the milk carton. If you have a mirror with a magnifying side and a nonmagnifying side, have the nonmagnifying side facing the hole.

To make a periscope from a 1-quart milk carton, your mirrors must be smaller than 3¹/₂ inches in at least one dimension. If the only mirrors you can find are larger than that, you can use half-gallon milk cartons instead.

What if I want to use half-gallon milk cartons or some other boxes?

When you are making a periscope, it's important to make sure that your mirror is positioned at a 45-degree angle. If you use a wider milk carton or some other box, just measure how wide your box is. Then measure that

same distance up the side of the box and make a mark. The line between your mark and the opposite corner of the box will be at 45 degrees.

How does my periscope work?

Light always reflects away from a mirror at the same angle that it hits the mirror. In your periscope, light hits the top mirror at a 45-degree angle and reflects away at the same angle, which bounces it down to the bottom mirror. That reflected light hits the second mirror at a 45-degree angle and reflects away at the same angle, right into your eye.

Can I make a periscope with a really long tube?

You can make your periscope longer, but the longer the tube is, the smaller the image you'll see. Periscopes in tanks and submarines have magnifying lenses between the mirrors to make the reflected image bigger.

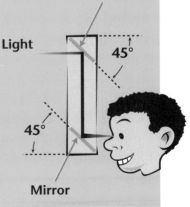

Looking for Looking Glasses

Next time you're in a restaurant and the kids are bored, polish your spoon on a napkin to make a great funhouse mirror!

You can see your face reflected in a mirror—and you can see your reflection in other places, too. Many smooth and shiny surfaces can act like mirrors. Curved surfaces act like weird mirrors, twisting and bending your reflection to make you look stretched out of shape and strange.

Search your house for smooth and shiny things that act like mirrors. You can start by looking at yourself in a shiny metal spoon. First, hold a spoon so that it's straight up and down. Then look at the back of the spoon. You'll see a reflection of your face that's stretched so that it's long and skinny. If you turn the spoon sideways you'll see a reflection that's short and wide. Flip the spoon over and look into the bowl of the spoon. Suddenly, your reflection is upside-down! Watch what happens when you move the spoon up and down or from side to side.

After you play with a spoon, look for your reflection in a TV screen (when the TV is turned off), glass bottles, shiny metal pots, chrome faucets, and other people's eyes.

Once you find your reflection in something, see if moving around changes how your reflection looks. In a cooking pot, your face will look skinny if you hold the pot one way and wide if you hold the pot the other way. If you have trouble finding reflections, read "Tips for Reflection Hunters," on the following page.

The Science-at-Home Team's Field Trip

When the Science-at-Home Team started looking for looking glasses, we found them in all sorts of interesting places. Ellen met a friend for dinner in a tea shop. She and her friend found their reflections in teapots and teacups, in the pictures on the walls, and in the silverware. She said they startled the other customers by waving their hands to find their reflections in a large silver urn across the room.

Pat went for a walk after work. She saw her reflection in the still waters of the lagoon in the city park next to the Exploratorium. When a duck swam past, the ripples disrupted her reflection.

Linda found her reflection in the shiny plastic of a full bottle of cola. But when she drank up all the soda, she couldn't see her reflection in the bottle anymore. She

Tips for Reflection Hunters

Sometimes, it's tough to spot a reflection. Here are some ways to make finding reflections easier.

Keep moving. You're more likely to see something that's moving than something that's standing still. Move your head around or hold up your hand and wiggle your fingers. Look for movement in the reflector.

Wear white or brightly colored clothes. You see your reflection when light bounces off you, then bounces again off something smooth and shiny, and finally gets into your eyes. When you're wearing white, more light bounces off your clothes, so there's more light bouncing off the reflector and getting into your eyes. Or you could try wearing a bright color, like a red shirt. Then, if you see a big patch of red in the reflection, you'll know that's your shirt.

Pay attention to the light. If you look for your reflection in something clear (like a glass bottle), light shining through the bottle can make it hard to see the light reflecting off the bottle. Putting a dark piece of paper behind the bottle will block out the light and make it easier to see your reflection.

explained that the light shining through the empty bottle washed out her reflection. When the bottle was full, the brown cola blocked the light and made it easier to see her reflection.

Jenefer (a teacher who worked with the Science-at-Home Team) found great reflections in her kitchen faucet, and on lots of metal pots.

That's what we found on our search for looking glasses. How many weird mirrors can you and your family find?

What's Going On?

Why do different reflectors stretch my face in different ways?

You see your reflection in a mirror because the mirror bounces light back into your eyes. Your eyes and brain use that light to make a picture of the world.

When your brain makes a picture of the world, it assumes that all the light getting into your eye has traveled in a straight line to get there. But when you see a reflected image, that light didn't travel in a straight line to your eyes—it bounced off a mirror. But your brain makes a visual picture of what you would see if the light had traveled in a straight line from your face to your eyes. So you see your face in the mirror.

Things get a little tricky when you look at your reflection on a curved surface—like the back of a spoon (a *convex* surface). A curved surface sends light bouncing away at a different angle than a flat surface would, distorting your image. What you see in the reflection depends on what direction the light is coming from when it gets into your eyes.

The bowl of a spoon (a *concave* surface) reflects light from the top of your head so that it comes down toward your chin. It reflects light from your chin so that it comes up toward the top of your head. When your brain interprets this light, you see a reflection of your head that's upside-down!

Surprising Structures

2

If you wanted to build a tower that was 6 feet wide and 20 feet tall, what would you use? Wood and nails? Cement? Bricks? You'd probably figure that the tower should be built out of something very strong, or it would just fall down. And you'd be absolutely right.

At the Exploratorium, we used newspaper.

In the spring of 1994, Vivian Altmann and Ken Finn from the Exploratorium's Children's Outreach Program, and Jok Church, creator of "Beakman's World," held a Scale and Structure Weekend. Their goal was to see how tall a tower they could build out of newspaper and tape. They started building it under a skylight in the middle of the museum, so most of the visitors who were there that day joined in the fun.

The first problem was how to turn newspaper into a strong building material. A single sheet of newspaper is flat and floppy. Even a baby could rip it in half. But Viv, Ken, and Jok soon discovered that the same piece of paper became much stronger when it was rolled up into a long, tight "stick."

They taped 3 sticks together into a triangle, then added more triangles onto it to make a structure called a *tetrahedron.* Each tetrahedron stood up on its own, and 6 or 7 of them taped together made a base for the

tower. They built up 3 layers—about 6 feet—of tetrahedrons before they ran into another problem: you can't climb up a newspaper tower to add more layers. The newspaper "sticks" bend in the middle if you step on them.

So they built a new layer on the floor. Then Viv called everyone over, and all together they lifted up the whole tower, and the new layer went on the bottom! By Sunday afternoon, the tower was more than 20 feet tall.

In this chapter, you can experiment with some things around your house that make weird and wonderful building materials. Make your own towers out of toothpicks and gumdrops, and build forts and walls out of blocks of sugar. You'll also discover some surprising ways to cut and fold structures out of paper, and go on a field trip through your own neighborhood to see how things are built.

Paper Architecture

Turn flat pieces of paper into 3-dimensional art!

What Do I Need?

- A package of lined file cards
- Scissors
- A pencil
- A ruler

What Do I Do?

Making a Pop-Up Shelf

1 Fold a file card in half like this:

From the folded side, draw 2 lines about 1 inch apart. Make each line about 1 inch long.

2 Use your scissors to cut the 2 lines, starting from the fold. You've made a flap.

3 Fold the flap forward. Crease it with your fingers. Now fold it back the other way, crease it again, then unfold it.

4 Open up the file card. Put your finger on the back of the card and push the cut part out toward the front. The whole card will be bent in the middle. When the cut part pops out, it will look like a little shelf.

5 Use your fingers to crease the edge of the shelf so it stands up straight. Carefully close the file card again and press down to make all the creases stronger.

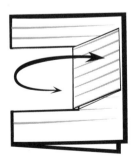

Making Spectacular Stair Steps

1 Fold a file card in half like this:

Use the lines of the file card as a guide to draw pencil lines up from the fold. The two shortest lines will be on the outside edges of the card. They should be about 1/4 inch high. As you move in toward the center, each pair of lines will be about 1/4 inch taller. The last two lines (in the very center) will be about 1 1/4 inches high.

2 Cut on each vertical line. You'll have 9 narrow flaps. Carefully fold each flap up and crease it at its folded edge. Unfold all the flaps.

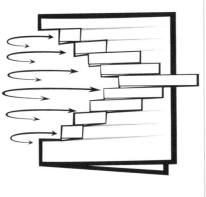

3 Open the file card. Use your fingers to pop each stair step out, starting from the biggest one in the middle. Pinch each crease between your thumb and finger. (This is tricky and might take a little practice.)

4 When all the stair steps are popped out, carefully close the file card and crease it tightly with your fingers.

To make any of your finished pop-ups stronger, fold a second file card in half and glue it to the back and bottom of the first card.

What Else Can I Try?
Here are some experiments you might want to try.
- Can you make 2 shelves side by side?
- What would your shelf look like if you cut curved lines instead of straight lines? How about if you cut one line longer than the other and folded the flap at an angle?
- What would happen to the mouth pop-up (see page 24) if you cut the line on one side of the card instead of in the center?
- What would the stair steps look like if the tallest flaps were on the sides instead of in the middle?
- What interesting new shapes can you create by experimenting on your own?

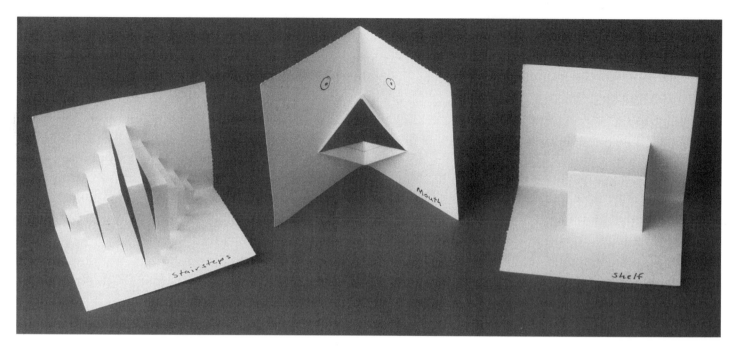

Making a Pop-Up Mouth

1 Fold a file card in half like this:

Use your scissors to cut from the center of the folded side into the middle of the card.

2 Fold a triangle out from the cut to one edge of the card. Crease the edge, then fold it back the other way and crease it. Repeat for the other side of the cut.

Eleven-year-old Anna Morris made a file-card monster with teeth. When she made it eat her nose, it hurt a lot more than she thought it would!

3 Unfold both triangles so the file card is flat. Open the card like a book. Put your finger on the back of the card and push the triangles toward the front, letting the card fold shut a little as you push. When you open and shut the card, you can make the mouth in the middle look like it's talking.

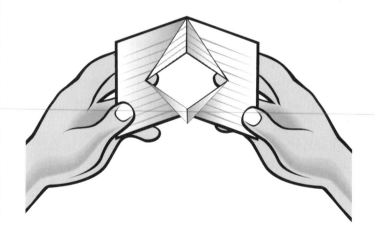

4 If you want, you can draw eyes and lips on the mouth, or make a word balloon. (This is a fun pop-up to use for party invitations.)

Making Stacked-Up Shelves

1 Make your basic pop-up shelf (see page 22). Pop the shelf out, crease it, and then fold the card flat.

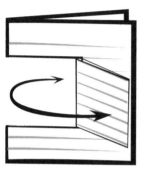

2 Draw 2 parallel lines on the shelf you just made.

Cut on those lines to make a new shelf. Fold the thick flap forward, then back. If you want, you can cut a third shelf into the new one.

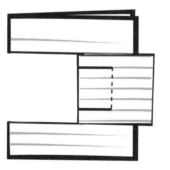

3 Open the file card and pop out all the shelves. One will look like it's stacked on top of the original shelf; the other will jut out below.

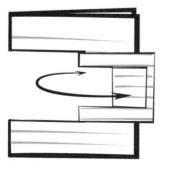

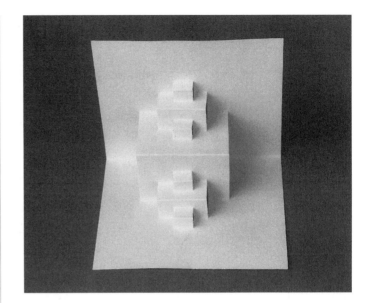

4 You can continue making generations of shelves by refolding the card and cutting more flaps. Each generation will make twice as many shelves as the one before, and each shelf will be about half the size. After 2 or 3 generations, this can be very tricky, because you're cutting, folding, and popping out very tiny flaps from very thick layers of file card.

What's Going On?

What do pop-ups have to do with building things?

When you made your first pop-up, you may have been surprised that such simple cuts could make something so unexpected. Even after experimenting for a while, you may find that it's hard to predict what will happen when you make a certain fold or a certain cut.

Engineers and architects need to develop an ability to visualize how shapes fit together in three dimensions. Most people never take the time to exercise that part of their brain. Many of us have trouble visualizing shapes in three dimensions and figuring out how they fit together. Experimenting with paper pop-ups and *origami*—the art of folding paper—is one way to exercise your visual imagination.

Build a Box

Make a sturdy box from a single piece of paper.

What Do I Need?

- A rectangular piece of paper (You can make these boxes out of paper of any size, as long as it's a rectangle. Copier paper or notebook paper works fine. Pages cut from magazines make very colorful boxes.)

What Do I Do?

1 Fold the paper in half the long way. Unfold the paper. After each fold, use your fingers to make a sharp crease.

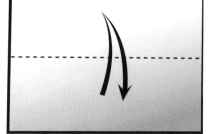

2 Fold the long edges in to the center crease you just made. Unfold the paper.

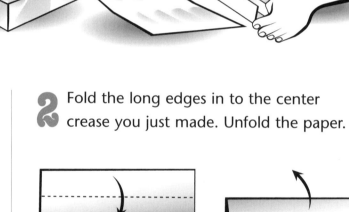

3 Fold the paper in half the short way. Unfold the paper.

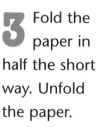

4 Fold the outer edges in to the center crease you just made. DON'T unfold the paper.

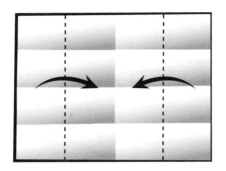

5 Now you have a rectangle with 2 cut edges running down the middle and 3 creases going across. Fold each outside corner until it touches the first crease near it.

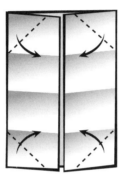

6 There are 2 strips in the center that are not covered by corners. Fold them over so they cover the tops of the corners.

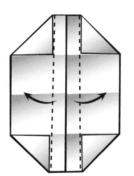

7 Slide your fingers under the middle of each of the 2 narrow folded strips. Gently pull out to form the box. Crease all the edges and corners with your fingers.

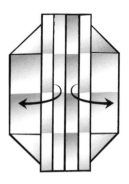

8 If you want a box with a lid, follow steps 1–7 again with a second sheet of paper. If you're using copier paper or magazine pages, the box and the lid can be the same size. Just tuck the box into the lid. If you're using stiffer paper, make the lid first, and cut the second piece of paper 1/4 inch smaller all around to make the box itself.

For a sturdier box, cut a piece of cardboard that's the same size as the bottom of the finished box, and put it inside.

Wow! I Didn't Know That!

The Japanese use paper in almost every part of their lives. There are paper walls and windows in Japanese houses. People carry waterproofed paper umbrellas in the rain and light their gardens with paper lanterns. Some people spend their whole lives learning to master the arts of *origami* (paper folding) and *kirigami* (paper cutting).

Your box will also float! You can use it with the other bathtub experiments in the "Sink or Swim" chapter on page 99.

File Card Bridges

How many pennies will your bridge hold?

What Do I Need?

- 4 to 6 books (enough to make 2 stacks the same height)
- A package of file cards
- 300 to 400 pennies (loose or in rolls)
- Scissors

What Do I Do?

1 Make 2 stacks of books with a gap of about 4 inches between them. Make sure the stacks are the same height.

2 Lay one file card over the gap between the books. About 1/2 inch of the card should be resting on a book at each end. How many pennies do you think you can pile on this flat bridge before it falls into the gap— 5? 10? 100? Try it and see how close your guess was.

Wow! I Didn't Know That!
As an advertising stunt for a paper company, Lev Zetlin Associates designed a full-sized paper bridge that was strong enough to support a car!

3 Without adding anything to the file card, try to make your bridge stronger. How could you change a file card to make it stiffer? What happens if you fold the card in half? If you make an arch? How about if you fold the card into pleats?

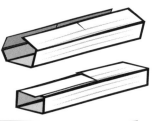

4 Make a bridge, then test it to see how many pennies it will hold.

Some of your bridges may hold a few pennies before falling down. Others may be stronger, but the pennies may slide right off. And some bridges will probably hold a lot more pennies than you'd think.

What's Going On?

How many pennies can my file-card bridge hold?

You may find that a file-card bridge can hold more pennies than you'd think! Here are the results of the file-card bridges that the Science-at-Home Team built.

Paper arch: 40 pennies

Flat card: 15 pennies

Folded beam: 175 pennies in rolls

Corrugated (pleated) bridge: 400 pennies (about 2 1/3 pounds!) in rolls.

A roll of 50 pennies weighs 132 grams—that's a little more than 4 1/2 ounces.

How many kinds of bridges are there?

You might think that bridges come in an infinite variety of forms. But if you get right down to the structural elements of a bridge, there are really only three kinds: beam spans, arch spans, and suspension spans.

The simplest kind of bridge is a beam bridge. A log that has fallen across a river makes a beam bridge. So does a board laid across a puddle, or a span of steel laid across a body of water, or a file card laid across two books. A beam bridge relies on the stiffness of the building material. If the log across the river sags, it doesn't make a very good bridge.

Arches have been common features in buildings since 1,000 B.C., but they didn't appear in bridges for another thousand years. Roman roads, built at the height of the Roman Empire's power, were often supported by stone arches.

Suspension bridges, like the Golden Gate Bridge in San Francisco, rely on a cable or rope for their support. Each end of the cable or rope must be anchored to the bank—tied to a tree, a boulder, or (in modern suspension bridges) a massive block of concrete called an anchorage. The cable or rope pulls on the anchors, but as long as they don't move and the cable or rope doesn't snap, the bridge is stable.

What kinds of bridges can I make with my file cards?

Using just your file card, you can make two of the three different kinds of bridges. When you lay a file card across two books—even if you've folded the card into pleats first—you've made a simple beam bridge. If you cut slots into the card, tuck the flaps under the edges of the book covers, and push the books slightly together, you'll make an arch bridge. We haven't figured out how to make a suspension bridge out of a file card, though. If you come up with a way to do it, please let us know!

> I cut slots in a file card and bent the flaps under the edges of my books to make a stronger file-card bridge.

Edible Architecture

Turn your kitchen into a construction site.

What Do I Need?

- A box of sugar cubes
- Waxed paper or plastic wrap
- A butter knife
- Sugar-cube mortar (recipe below)

Sugar-Cube Mortar

(makes about 1 1/2 cups)

- 3 egg whites
- A 1-pound box of confectioners' sugar
- 1/2 teaspoon cream of tartar
- An eggbeater
- A plastic container with a lid

Put the first 3 ingredients in a bowl. Beat them until they make a thick, white paste. Put the finished mortar in the plastic container so it doesn't dry out.

What Do I Do?

1 These building blocks can get very sticky. Cover your tabletop with waxed paper or plastic wrap.

This Is Important!

Sugar-cube mortar may look tasty, but it's not good to eat! Eggs should be cooked before you eat them.

2 Try making a tower out of just sugar cubes. Stack one cube on top of another. How high a tower can you build? One with 5 cubes? 10 cubes? 20 cubes?

3 Now try using some mortar with your cubes. Use a butter knife to spread a little bit of mortar on the top of a sugar cube. Stack another cube on top of it. Keep stacking cubes and mortar until your tower topples over. How tall did it get?

4 To make a bigger tower, start with a base of 4 cubes. Use the mortar to stick 4 cubes together. You'll end up with a square made of 4 sugar cubes. The insides of the square are all stuck together, and the outsides are dry.

5 Spread some mortar over the top of the square. Repeat step 4. Keep adding as many layers of cubes and mortar as you can.

The Bath/Plewa family discovered that if you let the mortar get thick (like cement) between layers of sugar cubes, it lets you build really strong, tall structures. Home Scientists in the Clear family didn't want to waste the leftover egg yolks from their sugar-cube mortar, so they made a sponge cake. What a great idea!

Mortar glues the cubes together so they don't wiggle as much, and so your tower won't fall over as easily. Can you make a 10-layer tower? A 20-layer tower? Taller? Can you build a pyramid with cubes and mortar? What else can you build?

What's Going On?

How can I make my tower more stable?

The bigger the base of your tower, the taller you can build. A pyramid with a base of nine sugar cubes is less likely to tip over than a tower with a base of just four sugar cubes.

When you build, you're experimenting with gravity and balance. You build your tower up, and gravity tries to pull it down. To remain upright, your tower has to be balanced on its base. To keep your tower balanced, you need to think about how its weight is distributed.

Physicists talk about *center of gravity*. To get a feel for center of gravity, imagine balancing a yardstick horizontally on your finger. When your finger is at the center, the yardstick balances. Now suppose you tape some pennies to one end of the yardstick. To balance the yardstick again, you have to move your finger closer to the weighted end of the stick. Adding weight to only one side of the yardstick moves the center of gravity. When something is balanced, its center of gravity is over the balance point.

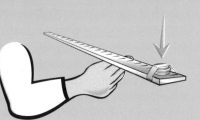

A tall, thin tower of sugar cubes will eventually start to lean. It's inevitable—sugar cubes aren't manufactured to precise specifications and their sides aren't always parallel. When a thin tower leans enough, a line drawn straight down from its center of gravity falls outside the tower's base. Then there's nothing to support the center of gravity and the tower tips over. If the tower has a broad base, it can lean a little—and its center of mass will still be over the base. That's why a wide base makes a tower more stable.

Why aren't bricks shaped like sugar cubes?

Next time you see a brick wall, notice that the bricks are arranged so that they overlap. There's no vertical line of mortar that runs straight through the whole wall. The mortared places where bricks meet form weak points where the wall could easily split. Since bricks are rectangular (rather than square, like sugar cubes), it's 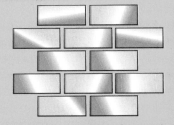 easier to create an overlapping arrangement where the weak points don't line up.

Geodesic Gumdrops

Make amazing architecture with candy and toothpicks.

What Do I Need?

- A bag of gumdrops (If you can't find gumdrops, try using bits of rolled-up clay, mini-marshmallows, or partly-cooked beans. Be creative!)
- A box of round toothpicks

What Do I Do?
Making Squares and Cubes

1 Start with 4 toothpicks and 4 gumdrops. Poke the toothpicks into the gumdrops to make a square with a gumdrop at each corner.

2 Poke another toothpick into the top of each gumdrop. Put a gumdrop on the top of each toothpick. Connect the gumdrops with toothpicks to make a cube. (A cube has a square on each side. It takes 8 gumdrops and 12 toothpicks.)

3 Use more toothpicks and gumdrops to keep building squares onto the sides of the cube. When your structure is about 6 inches tall or wide, try wiggling it from side to side. Does it feel solid, or does it feel kind of shaky?

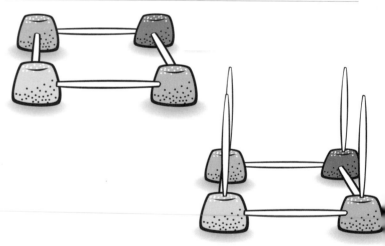

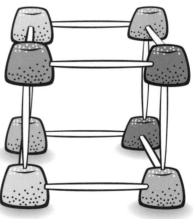

Making Triangles and Pyramids

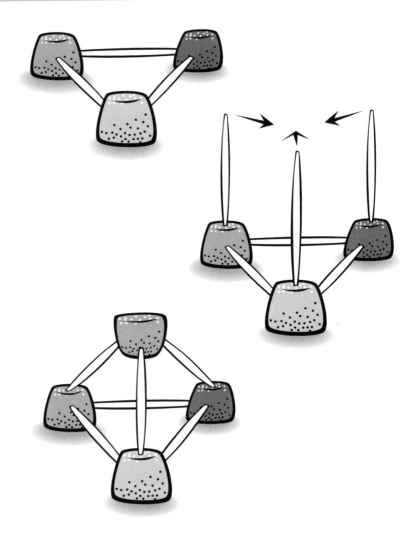

1 Start with 3 gumdrops and 3 toothpicks. Poke the toothpicks into the gumdrops to make a triangle with a gumdrop at each point.

2 Poke another toothpick into the top of each gumdrop. Bend those 3 toothpicks in toward the center. Poke all 3 toothpicks into one gumdrop to make a 3-sided pyramid. (A 3-sided pyramid has a triangle on each side. It takes 4 gumdrops and 6 toothpicks.)

3 Use more toothpicks and gumdrops to keep building triangles onto the sides of your pyramid. When your structure is about 6 inches tall or wide, try wiggling it from side to side. Does it feel solid, or does it feel kind of shaky?

Wow! I Didn't Know That!

Some of the structures you made out of gumdrops and toothpicks may look a lot like the models made by Buckminster Fuller. He was an engineer who experimented with strong, lightweight structures. He formed triangles out of thin aluminum rods and kept adding more triangles until he'd made an *icosahedron*—a globe made of 20 equal triangles. Fuller called it a "geodesic dome." Epcot Center at Disney World is probably the most famous geodesic dome.

The Frost family built a toothpick-and-gumdrop church with a tall steeple and a covered arch leading to a smaller building. The Optow-Chang family used bread cubes instead of gumdrops, and then made croutons from the leftovers. The Goers family built a gumdrop house with a staircase inside. Then they ate it.

Making 4-Sided Pyramids

You can make a very big structure out of squares and cubes, but it'll be wiggly and will probably fall down. If you try to make a structure out of only triangles and pyramids, it won't be wiggly, but you'll probably run out of gumdrops and toothpicks before it gets very big. A 4-sided pyramid has a square on the bottom and triangles on all 4 sides. When you make a structure that uses both triangles and squares, you can make big structures that are less wiggly.

1 Build a square, then poke a toothpick into the top of each corner.

2 Bend all 4 toothpicks into the center and connect them with one gumdrop, to make a 4-sided pyramid.

3 What other ways can you use squares and triangles together? How big a structure can you make before you run out of gumdrops?

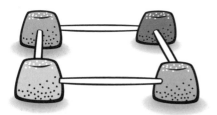
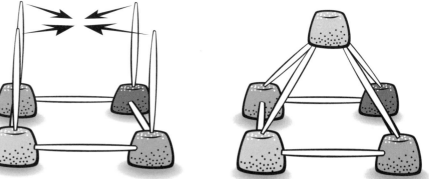

What's Going On?

Stretching and squashing—some basic principles

Even though your gumdrop structures are standing absolutely still, their parts are always pulling and pushing on each other. Structures remain standing because some parts are being pulled or stretched and other parts are being pushed or squashed. The parts that are being pulled are in *tension*. The parts that are being squashed are in *compression*.

Sometimes you can figure out whether something is in tension or compression by imagining yourself in that object's place. If you're a brick and someone piles more bricks on you, you'll feel squashed—you're in compression. If you're a long steel cable attached to a couple of towers and someone hangs a bridge from you, you'll feel stretched—you're in tension.

Some materials—like bricks and sugar cubes (see Edible Architecture on page 30)—don't squash easily; they are strong in compression. Others—like steel cables or rubber bands—don't break when you stretch them; they are strong under tension. Still others—like steel bars or wooden toothpicks—are strong under both compression and tension.

What's the big deal about triangles?

As you've probably already discovered, squares collapse easily under compression. Four toothpicks joined in a square tend to collapse by giving way at their joints, their weakest points. A square can fold into a diamond, like this:

But if you make a toothpick triangle, the situation changes. The only way to change the angles of the triangle is by shortening one of the sides. So to make the triangle collapse you would have to push hard enough to break one of the toothpicks.

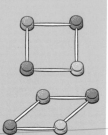

If you want to, you can use your gumdrops and toothpicks to build some strong structures that are made by combining triangles and squares. Here's a pattern that's similar to some used in modern bridge design:

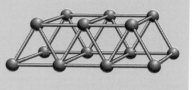

The Triangle Treasure Hunt (page 35) may give you ideas for other designs you can build with gumdrops and toothpicks.

Triangle Treasure Hunt

When you built towers from gumdrops and toothpicks, you used both squares and triangles—and you probably discovered that 4 toothpicks joined in a square tended to wiggle and fold into a diamond shape. But when you made triangles, they didn't fold up. Triangles let you build stronger structures than squares.

When you look at buildings, it's easy to find squares and rectangles. But if you look closely, you'll find triangles as well—especially in structures where you can see the supporting framework. For the next few days, look for triangles in the world around you: in your house, in your neighborhood, in the pictures you see in magazines or the shows you see on TV. You might want to keep a list of triangles that you see or draw pictures of some of the places you saw triangles. Think about why there are so many triangles around you.

The Science-at-Home Team's Field Trip

The Science-at-Home Team went looking for triangles in and around the Exploratorium. Here are some of the things that we found.

Hanging in the closet were coat hangers, perfect triangles with a hook to hang them up. The sides of the triangle hold the shoulders of a coat, and the bottom bar keeps the sides of the triangle from squishing together. Now that you know about tension and compression, you know that the sides of the triangle are under tension and the bottom bar is under compression.

Some of the shelves in the Editorial Department office are supported with brackets that are triangular. They look like this:

In the computer room, there are steel shelves made from many steel bars. From the side, they look like lines of triangles.

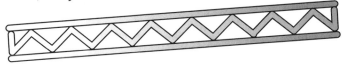

The Exploratorium is in a building constructed for the 1915 Panama-Pacific International Exposition. From the exhibit floor, we looked up at the ceiling and saw girders that looked like this:

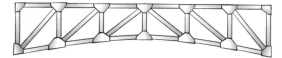

When we went outside, we noticed a nearby house that was under construction. The workers had put up scaffolding that had many triangles. It looked like this:

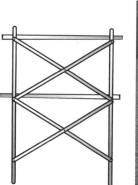

As we walked down the street toward San Francisco Bay, we saw triangular roof lines. When we reached the seawall, we looked across the water at the Golden Gate Bridge, a collection of many triangles. Take a look at the photograph of the bridge and see how many triangles you can find.

Those are the triangles that we found in just an hour of looking. What can you find?

What's Going On?

Why triangles?

A square can collapse a little by giving at the joints. To collapse a triangle, you need to shorten one or more of its sides.

Why aren't there more triangles?

If you see a house under construction, you'll see lots of rectangles. The wall of a house that's going up often looks something like the picture to the right.

Lots of rectangles, but you may not see many triangles. But before the house is done, the construction workers nail sheets of plywood to the outside. The plywood connects all the vertical and horizontal boards. Like a

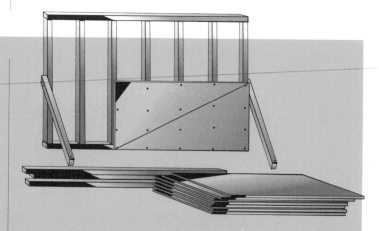

triangular brace, the rigid plywood keeps the rectangles from collapsing.

Of course, if you look at the wall from the side, you're likely to see a brace holding the wall up. The brace, the wall, and the ground will make—what else—a triangle!

Taking Things Apart

One way to figure out how something works is to look inside it. At the Exploratorium, we like to take a close look at the inner workings of all kinds of things. Karen Kalumuck, one of the Exploratorium's biologists, likes to cut open fruits and vegetables and think about how they grew. Linda Shore, one of the Exploratorium's physicists, likes to help people learn how to take apart mechanical things— like floppy disks and wind-up toys.

Linda and Karen both have advice for anyone who wants to take things apart. No matter what you're taking apart, the first step is to look at it really carefully. What does this thing do? What do you notice about it?

Karen likes to write down what she sees when she looks at something she is about to take apart. Or sometimes she draws pictures. After something's in pieces, you may need a picture to remember what it looked like to start with.

When you start taking something apart, Linda says, it's important to take your time. Don't rush, or you may miss something interesting.

When you are taking apart mechanical things, be sure to save all the pieces. As you take off each piece, look to see how it connects to the other pieces. Could you put that piece back on if you wanted to? If the thing that you're taking apart doesn't work, see if you can figure out why. Is there one piece that's broken? Are there two pieces that used to connect, but don't anymore?

Maybe you can't fix the thing that you're taking apart. But you can at least see how cleverly it was put together. And that may help you figure out how to make something clever yourself. In this chapter, we give you instructions on how to take apart some simple things (like envelopes and bananas) and some more complicated things (like floppy disks and wind-up toys). We also suggest places you might look to find other interesting things to take apart. Have fun!

Envelope Encounters

Recycle your mail to discover some interesting shapes.

What Do I Need?

- A few different-sized envelopes
- A brown paper bag
- A pencil
- A marker
- Scissors
- Tape or a glue stick
- A newspaper
- Wrapping paper (if you want)

What Do I Do?

1 Look at an envelope. What shape is it—square or rectangular? Guess what shape it will be when you unfold it.

2 Find the places where the flaps of the envelope are glued together. Slide your finger under each flap and very carefully pull them apart.

3 Unfold the envelope and lay it out flat. What shape is it now? Is it the shape you guessed it would be?

4 Unfold some of your other envelopes. Are they all the same shape? How many different shapes can you find?

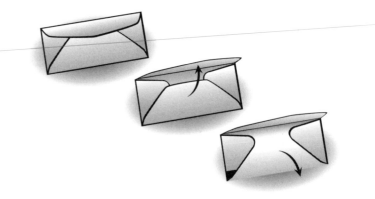

Making Your Own Envelopes

1 Unfold a brown paper bag to make a big, flat sheet of paper. Put that big, flat sheet on top of the newspaper. Lay one of your unfolded envelopes on top of the brown paper. Use a marker to trace around the outside of the envelope. Use the pencil to draw along the lines where the envelope folds. Press hard!

2 Cut out the envelope shape along the marker lines. You should be able to see faint dented lines where the pencil was pressing. Fold your new envelope on those lines. Fold the bottom (A) in first, then the two sides (B).

3 Tape or glue the bottom and the sides of the envelope shut. Fold the top flap down, but don't tape it until you put a letter inside.

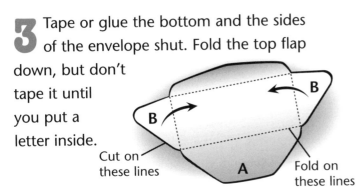

Cut on these lines

Fold on these lines

B   B   A

4 You can make very colorful envelopes from wrapping paper. If you want to mail a letter in a wrapping-paper envelope, make sure the post office can read the address. Tape a square of white paper to the front of the envelope and write on that.

Wow! I Didn't Know That!

The first company to make envelopes in the United States was started in New York in 1839 by a man named Pierson. Before that, people just folded up their letters. They wrote the address on the outside of the letter, and sealed the edges with blobs of sealing wax.

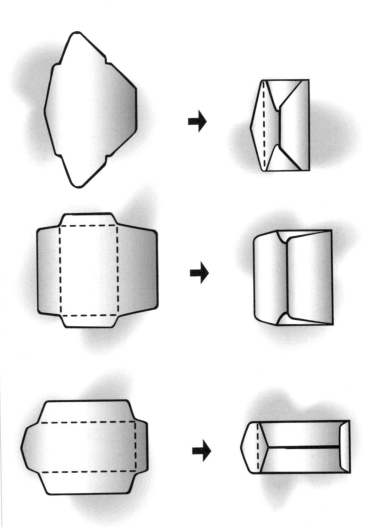

What's Going On?

Why bother to take apart envelopes?

At the Exploratorium, we like to look closely at familiar things. Sometimes, a closer look makes ordinary things seem extraordinary!

An ordinary envelope doesn't look like much—it's just a rectangle with a flap on top. But when you take an envelope apart and flatten it out, its shape may surprise you.

By taking an envelope apart, you also create a pattern that lets you make more envelopes, just like the first one.

What else can I do?

Our modern society is filled with packaging that's designed to hold something while it's in the store and then get thrown away. Before you recycle your paper packaging, take it apart and see how it's put together.

Dissect a Disk

Find out what's inside a floppy disk.

What Do I Need?

- A 3.5-inch computer disk
- A butter knife

What Do I Do?

1 Ask a grown-up for a computer disk that you can take apart. DON'T just take one! It might have important stuff on it.

2 Pretend you've never seen a disk before, and look at the outside of it carefully. What different parts can you find? Do any of them move? Does anything change when you move those parts?

3 Now look for the places where the disk will come apart. You can take it apart any way you want, but an easy way to start is to use a butter knife to carefully pry up the metal rectangle—the shutter—that folds over one edge of the disk. Pull it off and put it aside.

4 Now you can see some slots in the plastic. Put the knife in a slot and gently pry apart the two flat halves of the disk. Don't just snap the disk apart, or you'll lose the tiny spring inside. It held the shutter closed and may jump out, so watch for it. The plastic at the corners of the disk will probably crack or break off. That's okay.

5 Do the two halves of the disk look the same? What parts can you see? You may want to draw a picture of where everything is before you take the disk apart any further.

6 Here are all the parts of the disk that you can find if you look carefully:

- A metal shutter
- A spring
- A brown plastic circle from a metal center or hub (The circle is made out of the same stuff cassette tapes and videotapes are made of.)
- Two white paper rings
- A small black plastic rectangle with legs (Hint: it's up in one of the corners.)
- A small plastic flap
- Two plastic squares that hold everything else

Can you guess why each of these parts is there? How do you think they work together when the disk is inside the computer?

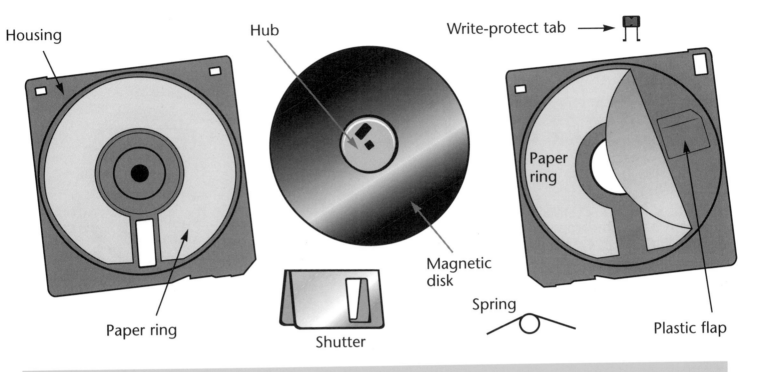

Housing • Hub • Write-protect tab

Paper ring

Magnetic disk

Shutter

Spring

Plastic flap

Paper ring

What's Going On?

What's inside a computer disk?

When you take apart a 3.5-inch disk, you'll end up with two colored plastic squares (the **housing**) that hold the other, smaller parts. Here's a guide to what each of those parts is, and what it does when the disk is inside your computer.

• **Shutter.** This is a piece of metal folded over one edge of the disk. That edge goes into the computer first. Inside the computer, the shutter slides over, and the information on the disk can be read through the rectangular slot.

• **Spring.** When the disk comes out of the machine, the spring snaps the shutter closed again so no dust or fingerprints can get onto the magnetic disk.

• **Magnetic disk.** This round piece of plastic is coated with iron oxide. Iron oxide can be magnetized. When you save information to a disk, a recording head creates a magnetic pattern on the iron oxide. The pattern stores your words or pictures in a form that the computer can read the next time you put the disk in.

• **Hub.** The metal center of the magnetic disk. The holes in the hub are like the hole in the middle of a record—they fit over spindles inside the computer and hold the disk in place while it spins.

• **Paper rings.** The magnetic disk is sandwiched between two white paper rings. The two rings are glued down to the plastic housing, and stay still while the disk spins. They clean the disk, removing microscopic bits of dust.

• **Write-protect tab.** This little plastic rectangle is in the upper right corner of most disks. It slides up to reveal a square hole in the housing (or slides down, to cover the hole). When the hole is open, the disk is locked. Your computer won't allow you to add anything to the disk or erase anything from it.

• **Plastic flap.** You have to hunt for this piece. It's tucked away under one of the paper rings. One end is glued down, and the plastic is bent, just a little. It functions as a simple spring that pushes the paper ring tight against the surface of the magnetic disk.

What's inside a cassette tape?

Now that you've taken apart a floppy disk, you may want to try dissecting an audiocassette or a videocassette. A lot of the parts on these cassettes are very similar to parts of a floppy disk. How many can you find?

• Magnetic tape.

• Hubs. There are two plastic wheels in the middle of a cassette. The tape winds from one to the other as they spin inside the machine.

• Write-protect tab. You can punch it out so no one can re-record over your favorite song or TV show.

• A piece of paper or felt (and a simple spring) that cleans the tape while it's moving.

• A shutter (and a spring) to keep dirt and fingers from getting to the tape. (Only on a videocassette.)

Wind-up Toys–The Inside Story

Take apart a wind-up toy and figure out what makes it work.

What Do I Need?

- A wind-up toy that you don't mind taking apart (you can use a broken one or one that works)
- A butter knife
- A screwdriver (Depending on the toy, you may need a regular screwdriver, a Phillips head screwdriver, or both.)

What Do I Do?

1 Take a really good look at your wind-up toy. What parts move when you wind it up? If it's broken, what did it used to do?

2 Most modern wind-up toys are made of pieces of molded plastic that fit tightly together. Sometimes, small screws hold the plastic parts together. First, look for screws. If you find some, take them out and see if you can take the toy apart.

3 If there aren't any screws, slide the tip of a butter knife or screwdriver into one of the seams where the plastic pieces fit together. Gently pry them apart. Work slowly and try not to break the plastic pieces. Keep all the pieces.

4 As you take the toy apart, notice how the moving pieces fit together with the rest of the toy. What pushes or pulls on a moving part to make it move?

5 At some point, you'll probably see a metal or plastic box, called a gearbox. The key or knob that winds the toy sticks out of this box. In some toys, at least one side of this box is open. If your toy is like this, take a close look inside before you go further. (If you can't see inside the gearbox, pry it open until you can.)

6 Inside the box, you'll probably see a spiral spring made from a flat strip of metal that's wound up into a coil. If your toy was working when you started taking it apart, wind it up now and watch what happens to the spring. When you wind the toy, you coil the spring tighter and tighter. By doing this, you are storing energy in the spring. It's that stored energy that makes the wind-up toy move.

7 What else is in the box? You'll probably see at least two gears of different sizes. Chances are the spring is connected to a big gear. When this big gear turns, it makes smaller gears turn. These smaller gears power the movement of the toy.

8 If you want, you can keep your gearbox in one piece and try to put the toy back together. Or you can try to pry the box apart and look at the gears up close. Be warned: prying the box apart can be difficult. And chances are you won't be able to put the toy back together once you take the gearbox apart.

9 If you want to keep going, think about what all those gears are doing inside the gearbox. The diagram below shows the gearbox from a wind-up dog that wags its tail. Here's what happens in that gearbox:

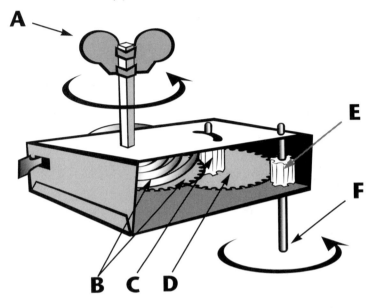

A Winding the toy makes the spring coil up tight.

B The spring uncoils slowly. As it uncoils, it turns the big gear that's attached to the spring.

C The big gear makes the little gear turn. Each time the big gear turns once, the little gear turns many times.

D Another big gear that's attached to the little gear turns.

E That big gear makes another little gear turn many times.

F That little gear makes the dog's tail spin around and around very quickly.

10 Is there anything interesting to do with the pieces you have left? We found that some of the gears we removed from the gearbox made great spinning tops.

Stuff to look for in your wind-up toy

What's going on in this wind-up pig?

Wind this pig up, and he walks. How do the parts inside the pig make that happen?

The gears inside the gearbox turn a cam. The cam fits into the round hole in the center of the pig. As the cam turns, it moves the cam follower back and forth. The cam follower is connected to the pig's legs by pin joints. As the cam follower moves, so do the legs.

Pin joint · Gearbox · Turning this will wind up the pig

Cam

Hole for cam · Cam follower

Most wind-up toys are filled with clever connections and mechanical devices. Here are some of the ones you may find in your wind-up toy:

Connections. You may see parts that are linked in a variety of ways. One common linkage is the *pin joint,* where a pin on one piece fits into a hole on another piece. When the part with the pin on it moves, the connecting piece moves too.

In the Watkins family, 8-year-old Tyler had a great time taking broken toys apart. Five-year-old Mallory couldn't see how all the pieces worked together very well, but it didn't take long for her to figure out how to use a screwdriver!

Wow! I Didn't Know That!

A wind-up clock has a spring that's a lot like the one in your wind-up toy. A clock is designed so that the energy stored in the spring will turn the clock hands at just the right speed.

Cam. You may see a circular piece that's connected by a pin joint to a long, skinny piece. That circular piece is a *cam.* If the connection is at the edge of the circle, the long, skinny piece (known as the *cam follower*) will move back and forth as the circle turns, changing the circular motion to a back-and-forth motion.

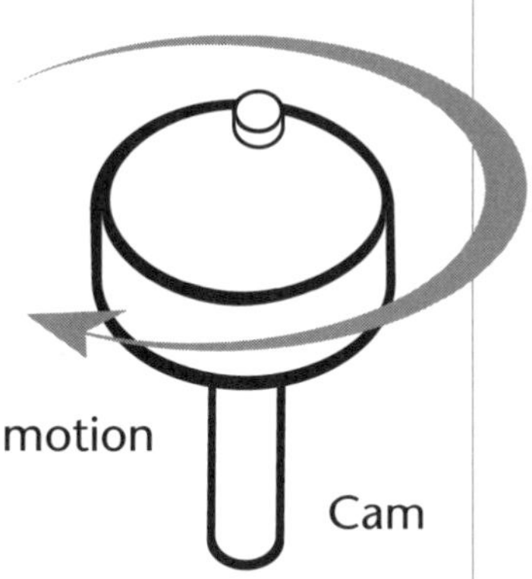

Cam

Constraints. Look for mechanisms that allow parts to move only in a particular way. In our swimming penguin toy, each leg had a slot with a pin through it. The slot let the legs move up and down, but they couldn't flop around in other directions.

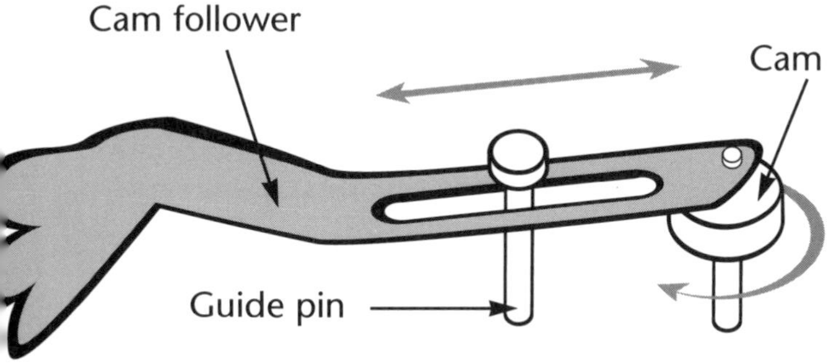

In wind-up toys, you can find mechanisms that change circular movements into back-and-forth movements. That's what's going on in this leg from a wind-up penguin.

Ratchet. A *ratchet* is a device that lets something spin in one direction, but not in the other. Here's one simple ratchet:

On many toys, a ratchet keeps the key from spinning backward immediately after you wind it.

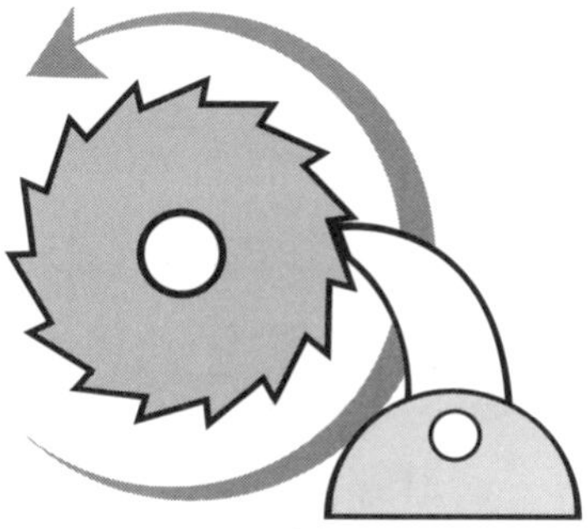

Ratchet

What's Going On?

Why bother taking apart a wind-up toy?

Taking apart a wind-up toy gives you a chance to look at some of the ways that you can transform one kind of motion into another. The spring turns a gear—and that gear turns other gears to make toys that walk, hop, waddle, and spin.

The mechanisms that you find in these toys are similar to those found in a variety of more practical devices, such as wind-up clocks, bicycles, and cars.

How long have wind-up toys been around?

The intricate workings of wind-up toys have a long and illustrious history. For centuries, people have been designing *automata,* mechanical objects that move on their own, often imitating live creatures. Back in 350 B.C. (or thereabouts), a friend of Plato's is said to have made a wooden model of a pigeon that made lifelike movements. The ancient Chinese made flying birds, an otter that caught fish, and moving human figures. In the Islamic world, inventors created mechanical peacocks that moved when water flowed through them.

In the middle of the fifteenth century, the coiled, tempered-steel spring was developed. Using this spring (the same sort of spring that's in your wind-up toy), people made intricate automata, like mechanical songbirds that popped out of snuffboxes and sang, and "magic boxes" where a mechanical magician pointed out the answer to your question.

Fruit Surprise

Are bananas and oranges alike? Cut them up and find out!

What Do I Need?

- An orange
- A banana
- A butter knife (if you want)

What Do I Do?

1 Think about an orange and a banana. How are they alike? How are they different? If you peel a banana and an orange, do they look alike inside? If you look closely, you may be surprised at what you find. What you see inside something can depend on just how you take it apart.

2 Look at the orange. If it's a navel orange, one end has a hole that looks like a belly button. (That's why they're called navel oranges. Navel is another word for belly button!) The other end has a little dry green thing. That's what's left of the stem.

3 Peel the thick skin off the orange with your fingers. You'll get a whitish ball that you can split apart into segments. Guess how many segments there are. Split the orange apart and see if you were right. (With some oranges you'll find seeds inside the segments, but navel oranges won't have seeds.)

4 Now look at the banana. It's got a short, dark stub at one end, and a stem at the other. The stem is where the banana was attached to the bunch, which hung upside-down from the tree.

5 Peel the thick skin off the banana. What's inside probably looks like it's all one piece, but it really isn't. Push your finger into the stem end of the banana. If you push very carefully, the banana will split into three long segments!

6 Run your fingernail down the center of one segment to find tiny dark spots. Those tiny spots are *ovules*. They look like seeds, but they can't grow into new plants.

7 If you have another banana  and orange, try cutting each one into round slices. Do they look alike that way?

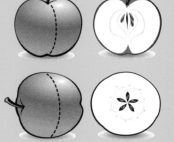

To a botanist, a banana and an orange are the same kind of fruit. They're both berries. (So are grapes and tomatoes.) Mangoes, peaches, and avocados are all stone fruits. They each have one huge seed. Apples and pears are core fruits. They both have a hard, crunchy center that's not good to eat. Cut up other fruits and see what they look like inside. You may find some surprises!

What's Going On?

What do bananas and oranges have in common?

All fruits start out as flowers. Different fruits develop from different parts of the flower. Seeds develop in the flower's ovary. As an orange flower becomes an orange, the whole ovary fills with juicy pulp. Seeds are scattered throughout this pulp. Bananas grow the same way. Botanists call this kind of fruit a berry. Oranges, tomatoes, bananas, and blueberries are all berries, according to botanists.

Why does a banana split so neatly into three segments?

Exploratorium staffer Kurt Feichtmeir showed us how to split a banana into three neat pieces. To find out why this works, we consulted Dr. W. John Kress, an expert on bananas and how they grow. He told us that the ovary of the banana flower has three sections, called *carpels*, where seeds can grow. When the ovary becomes a fruit, it still has those three carpels. When you push on the banana, you pry them apart.

Why can't I find the seeds in the orange? Are there seeds in the banana?

To produce seeds, a flower has to be fertilized. Navel oranges and the bananas that you buy in the store come from plants that are unusual—they produce a fruit even when the flowers aren't fertilized. Because the flowers weren't fertilized, these fruits don't have any seeds. To get more navel orange and banana trees, people grow new trees from parts of the existing plants.

Next time you eat a banana, look for the little black dots that are the ovules, or unfertilized seeds. In wild bananas, all those ovules develop into tiny, hard seeds, making the fruit much less pleasant to eat.

What else can I learn by making a fruit salad?

Next time you are making a fruit salad, cut one apple like this:

Cut another apple like this:

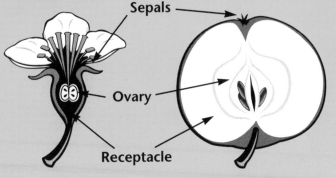

When you slice the apple from pole to pole, you get something that looks very different from what you get when you slice it around the equator. What you see when you look inside something often depends on how you took it apart.

The apple's core, the stiff stuff that surrounds its seeds, was the wall of the flower's ovary. The part of the apple that you eat was the receptacle, which is below the ovary. The dry, fuzzy stuff at the bottom of the apple used to be the sepals, the green leaves at the base of the flower.

Sepals

Ovary

Receptacle

Carrot Cut-up

Cut a carrot into thin slices and you can see beautiful patterns.

What Do I Need?

- A carrot
- A vegetable peeler
- A bright light

What Do I Do?

1 Take a close look at your carrot. Look for small roots that stick out of the carrot, or tiny "eyes" where roots used to be.

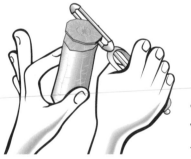

2 Wash a carrot and use the vegetable peeler to peel it. Ask a grown-up to cut about 2 inches off one end of your carrot. Use the vegetable peeler to slice off very thin rounds from the cut end. (This can be tricky, but after a few slices you'll get great thin rounds. Ask a grown-up if you need some help.)

Central cylinder

Cortex

3 Hold one of the rounds up to a bright light. Can you see the pattern of circles and lines?

What's Going On?

What can you see when you slice a carrot?

Carrots are roots. The root is the part of a plant that pulls water and minerals from the soil to feed the plant. Most roots absorb the water and dissolved minerals that a plant needs. A carrot does that—but it also stores food that the plant may need later.

The edible part of the carrot is the primary root, or tap root, of the carrot plant. When the carrot is in the ground, many small secondary roots grow from this tap root. If you look closely at the unpeeled carrot, you can see little eyes where the secondary roots were attached.

A cross-section of a carrot has a dark orange center surrounded by paler orange. Look closely and you'll see lines radiating from the center. If you slice the carrot lengthwise (a tricky maneuver that requires a sharp knife and grown-up assistance), you can see that the dark material that's in the center runs the length of the carrot.

The dark orange center, the central cylinder, contains tubes that carry water up from the roots to the leaves, and tubes that carry food from the leaves to the tip of the root. The paler orange, called the cortex, is where the carrot stores food. The lines that go to the central cylinder through the cortex are paths where water traveled from secondary roots to the central cylinder.

What Else Can You Take Apart?

By now you're probably getting pretty good at taking things apart. But even if you've done everything in this chapter, you're just getting started. Look around your house and see what other things you can take apart. They don't have to be complicated or mechanical.

If you do want to take mechanical things apart, though, there are a few dangers to watch out for.

- **Don't take apart anything that is electrical—anything that you plug in.** Even if an electrical appliance isn't plugged in, taking it apart can be dangerous. Some appliances have capacitors, which store electricity and could give you a fatal shock.

- **Don't take apart anything that has leaky batteries, glass that could break, or sharp metal parts that could cut you.**

If you follow these rules, you'll still find plenty of things that you can take apart using only a butter knife, a screwdriver, and a pair of scissors.

Whatever you decide to examine, remember to look at it closely and think about what it does. When you start taking it apart, don't rush; take your time. You may want to draw pictures or take notes to help you remember how it all went together. And save all the pieces so that you can see how they fit together when you're done.

The Science-at-Home Team Takes Things Apart

The Science-at-Home Team spent some time looking around our houses to see what we could find to dismantle. Then we got together to share the results.

Pat remembered taking apart wind-up alarm clocks when she was a kid, so she thought it would be fun to look inside a clock again. In the back of a drawer, she found a cheap wind-up clock. She used a small screwdriver to take out the screws that held the casing onto the clock face, unscrewed the wind-up key, and pulled off the casing, exposing the clock's inner workings. Inside, she found a spiral spring—very much like the one in a wind-up toy—that provided the energy to move the clock's hands. She also found the wheel that rocked back and forth with each tick of the clock and all the interlocking gears that made the hands move—the ones that moved when she turned the knob to set the time and the ones that moved a little bit with each tick of the clock, making the hands keep time. "Looking inside the clock made me appreciate how much like a wind-up toy it is," she said.

When Pat was a kid, she just put all the pieces of the clocks she took apart in a box and kept them. This time, she managed to put the clock back together. It seems to be keeping good time, but she won't trust it to wake her up for work.

Linda found an old flashlight in her garage and took that apart. A flashlight is a pretty simple machine. Electricity from the batteries makes a light bulb light up; a switch turns the bulb on and off. Linda knew that there had to be a complete electric circuit to make the bulb light up. The electricity had to flow from one end of the batteries, through the light bulb, and back to the other end of the batteries. When Linda unscrewed one end of the flashlight, she could see that one end of one battery touched the end of the light bulb. The other end of that battery touched the top of the second battery. The bottom of the second battery rested on a metal spring. A long thin piece of metal ran the length of the flashlight's tube, connecting the spring to the metal housing around the bulb and completing the electric circuit.

Ellen didn't want to dismantle anything she had at home, so she went out to search at garage sales and thrift stores. At one garage sale, she found a windup, walking Godzilla that looked just like a toy she had at home. When her Godzilla walked, sparks came out of his mouth. But the little monster she found at the garage sale didn't walk (or spark) any-more. When Ellen turned the key, it just made a whirring sound and then stopped.

Ellen took the broken Godzilla apart to see if she could figure out where the sparks

should come from. When she pried the toy's plastic body apart, she could see a tiny flint (like the one in a cigarette lighter) that rubbed against a metal piece attached to a gear. She turned the gear by hand and got sparks!

When we sat around and talked about it, we decided that we liked taking things apart because it helped us understand how they worked—and it also made us appreciate the clever things people had made. It was also fun to see how things that look very different on the outside can be a little bit the same on the inside. A sparking wind-up toy has a flint like a cigarette lighter and a wind-up clock has a spring like a wind-up toy.

That's what we discovered on our field trip. What did you discover on yours?

Our pioneering Home Scientists had a great time taking things apart. The Barton family took apart their broken bathroom scale—and fixed it in the process. The Schreyer family built a new toy with parts of the old ones they had dismantled. With parts from a flying bird and a sparking spinner, they made a kind of flying bird catapult! The Smith-Mancuso family showed their kids the inside of their dishwasher when they were repairing it. "Taking things apart and examining them is now a part of our lives," they told us.

Making Changes

The world is always changing. Metal rusts. Garbage decays. Ice and snow melt. Trees and flowers and molds grow.

Most people don't pay attention to the changes that are happening all around them. If they see a spot of fuzzy mold growing on a piece of bread, they just throw the bread away. But scientists and artists are very interested in changes—even changes like mold growing on bread.

Back in 1928, Sir Alexander Fleming was studying germs called *bacteria.* He filled dishes with food that bacteria liked to grow on. Bacteria grew in all the dishes except one. In that dish, there was a spot of mold. No bacteria grew near the mold—even though it grew everywhere else in the dish. Fleming discovered that the mold was making a chemical that kept the bacteria from growing. He called that chemical *penicillin*. Today, doctors use penicillin to help kill the bacteria that make people sick.

Antero Kare is a Finnish artist who uses mold in a very different way—he paints with it. Kare noticed that some molds looked like white powdery snow, and other molds looked like tiny green plants. In 1996, he built an Exploratorium exhibit called *Moosescape,* a model landscape made out of foam. Kare covered the hills and valleys with food that molds would grow on—mashed potatoes, boiled onions, oatmeal, beef bouillon, and agar (a gel used to grow molds in science

labs). He took tiny pieces of different kinds of molds, and "seeded" the landscape with them. Then he waited and watched what happened.

Some molds grew very quickly; others grew slowly. Within a few days, the whole landscape was covered with a cottony white mold. After a week or so, that mold began to die, and patches of green and blue molds began appearing. Eventually, the landscape was all covered in different shapes and colors of mold.

Scientists and artists both learn about the world around them by carefully observing what happens and how things change. In this chapter, you'll have a chance to observe a lot of different changes. You can watch how various kinds of mold grow by building your own Mold Terrarium. You can turn cream into butter, cover a steel nail with a coating of copper, and watch the amazing crystals that can grow from different kinds of salt!

I Can't Believe It's Butter!

Turn cream into butter in just a few minutes.

What Do I Need?

- A clear plastic container with a screw-on lid
- 1/2 pint of whipping cream
- A nickel
- Salt (if you want)
- A clean dish towel

What Do I Do?

1 Pour the cream into the container. The container should be no more than half full. If you want salted butter, add a few sprinkles of salt to the cream now.

2 Wash the nickel in hot soapy water. Rinse and dry it, then put it into the container with the cream. Screw on the lid tightly.

3 Shake the container. You can hear the nickel rattling around. As you work, the sound of the nickel clanking against the container will let you know how you're doing.

4 Keep shaking the container for about 15 minutes. You can shake it up and down, back and forth, or in any combination you want. If your arm gets tired, ask someone to help you.

Here's what will happen:

- **1–3 minutes.** The whipping cream will coat the inside of the container. You'll be able to hear the nickel clanking in the container and the liquid sloshing around.
- **3–6 minutes.** The cream has turned into thick whipped cream. You'll still hear the nickel, but no sloshing.

- **6–10 minutes.** It may feel like nothing's happening. That's okay. Keep shaking hard enough so that you can hear the nickel. You can probably see into the container again. Try slapping the jar against your palm to move the thick white stuff from one side to the other.

- **10–15 minutes.** All of a sudden, the thick white stuff will separate into a thin whitish liquid and a yellow glob. Keep shaking for another 30 seconds, then carefully pour off the liquid. What's left in the container is butter!

5 Spoon your butter onto the dish towel. Wrap the butter in the towel and squeeze it. More liquid will squeeze out of the butter.

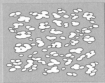

6 Spread your butter on some bread or toast. Or melt it and mix it with another amazing transformation—hard brown corn kernels that turn into fluffy popcorn.

Wow! I Didn't Know That!

Since the fifteenth century, Tibetan monks from the Gyuto Tantric Monastery have made beautiful sculptures out of tiny balls of colored yak butter. When the "butter monks" came to the Exploratorium in 1995, they had to use butter made from cows' milk because it's very hard to get yak butter in the United States.

What's Going On?

Why does shaking cream make butter?

Cream is a mixture of water (about 60%), fat (about 40%), and protein (just 1% or 2%). The fat is in tiny blobs that measure only about one ten-thousandth of an inch across. These blobs of fat are suspended in the water. They don't join to make one big blob of fat, because each tiny blob is coated with a protective membrane. When you shake cream, you break down the membranes that keep the blobs of milk fat apart, and the fat joins to make butter.

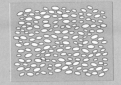

You've probably heard of people making butter from cream in a butter churn. Churning cream in a butter churn by agitating it with a wooden paddle is another way to break down the membranes that keep the fat blobs apart. In this activity, the nickel acts like the paddle in the butter churn.

Is there a scientific way of talking about what's going on here?

When tiny drops of one liquid are scattered throughout another liquid, that's what a chemist calls an *emulsion*. Cream is an emulsion of fat suspended in water. When you make butter, you've made a different emulsion. Tiny drops of water are scattered throughout your butter. So butter is an emulsion of water suspended in fat.

Crystal Creations I

Watch sponges and charcoal grow puffy "flowers" overnight.

What Do I Need?

- An aluminum pie pan, plastic food tray, large margarine tub, or other shallow container that you can throw away
- A paper cup or a glass jar that you can throw away
- Measuring spoons
- Toothpicks
- Charcoal, paper towels, and/or pieces of sponge
- Warm water
- Ammonia
- Salt
- Mrs. Stewart's Bluing (next to the bleach in the supermarket's laundry aisle)

Home Scientists in the Burns family liked the way green food coloring separated to make blue and yellow "flowers." The Perry family made many different crystal gardens. They glued toothpicks together to make houses and pyramids with paper-towel walls.

What Do I Do?

1 Have a grown-up put 3 tablespoons each of the salt, warm water, and bluing in the paper cup or jar. (Bluing stains just about everything, so be careful.) Stir until the salt is mostly dissolved. Add 3 tablespoons of ammonia, and stir again.

2 While a grown-up is mixing the chemicals, make the base for your garden. Arrange some charcoal, paper towels, and/or pieces of sponge in the bottom of a shallow container. Prop a few toothpicks up against them.

3 Have a grown-up carefully spoon all the chemical mixture over the charcoal, etc. Stir the solution before each spoonful so the salt in the bottom of the cup gets mixed in.

4 Put your garden in a safe, dry place and leave it overnight. A few white crystals may form in about 3 hours, but growth can be slow during the first day.

5 The next day, have a grown-up mix 2 tablespoons each of warm water, bluing, and salt. Spoon it into the bottom of the bowl around the base of the charcoal, etc. (Don't put the mixture right on the crystals or they will dissolve!)

6 If you add a tablespoon each of water, salt, and bluing to the bottom of the bowl every day, your garden should continue to grow for about a week. Each day you can also put a few drops of food coloring onto the charcoal, sponges, or towels so that your garden has a lot of different colors. (Remember not to splash any liquid on the crystals!)

What's Going On?

What if I can't find bluing in the store?

Mrs. Stewart's Bluing isn't in every store, but it's fairly common. If you can't find it in your neighborhood, you can call the folks who make the bluing at 1-800-325-7785 and order some. If that sounds like too much trouble, we've included a recipe for crystals made from Epsom salt (see Crystal Creations II on page 56), which is available in most supermarkets and drugstores.

Where did all those crystals in the crystal garden come from?

Pour a little table salt into your hand and take a close look at it. Table salt is made of many tiny crystals. When you mix these salt crystals with water, they dissolve, losing their crystalline form. When the water evaporates, the salt crystals form once again.

If you didn't add the bluing to the mixture, you would still get salt crystals when the water evaporated—but they wouldn't join to make crystal flowers. The bluing that you added contains tiny particles of an iron compound suspended in a liquid. These tiny particles provide places for the salt crystals to start forming, encouraging the formation of many linked crystals.

The crystals grow best where the liquid evaporates fastest. That's why the first crystals appear on the rim of the bowl and on the highest points of your garden.

Crystal Creations II

Grow spikes of crystals in the sun.

What Do I Need?

- Black construction paper
- Scissors
- A pie pan, cake pan, or shallow bowl
- Warm water
- Epsom salt (usually near the rubbing alcohol in the supermarket)

Tips for Home Scientists
This actvity works best on a sunny day.

What Do I Do?

1 Use your scissors to cut the black paper so it will fit in the bottom of your pie pan.

2 Add 1 tablespoon of Epsom salt to 1/4 cup of warm water. Stir until the salt is dissolved.

The Mudd family discovered that these crystals look great under a microscope.

2 Pour the salty water onto the black paper in the pie pan.

4 Put the pie pan out into the sun. When the water evaporates, you'll see lots of crystal spikes on the black paper.

What's Going On?

Why does Epsom salt make crystal spikes?

When you add Epsom salt to water, the salt dissolves. When you leave the pan in the sun, the water evaporates and the salt forms crystals shaped like long needles.

If you tried this experiment with table salt instead of Epsom salt, you wouldn't get crystal spikes. That's because table salt and Epsom salt are chemically different, so the crystals that they form are very different.

The picture on the right shows part of an artwork created for the Exploratorium by Swiss artist Jörg Lenzlinger. He mixed different kinds of salts with water. As the water evaporated, the salts crystallized, making beautiful shapes that kept growing and changing.

Copper Caper

Watch old pennies turn bright and shiny—right before your eyes!

What Do I Need?

- 20 dull, dirty pennies
- 1/4 cup white vinegar
- 1 teaspoon salt
- A clear, shallow bowl (not metal)
- 2 clean steel nails
- A clean steel screw or bolt
- Paper towels

What Do I Do?

1 Put the salt and vinegar in the bowl. Stir until the salt dissolves.

2 Dip one penny halfway into the liquid. Hold it there for about 10 seconds, then pull it out. What do you see?

3 Dump all the pennies into the liquid. You can watch them change for the first few seconds. After that you won't see anything happen.

4 After 5 minutes, take half of the pennies out of the liquid. Put them on a paper towel to dry.

5 Take the rest of the pennies out of the liquid. Rinse them really well under running water, and put them on a paper towel to dry. Write "rinsed" on the second paper towel.

6 Put a nail and a screw into the liquid. Lean another nail against the side of the bowl so that only part of it is in the liquid.

7 After 10 minutes, take a look at the nails. Are they a different color than they were before? Is the leaning nail 2 different colors? If not, leave the nails in the bowl and check on them again in an hour or so.

8 What's happening to the screw? You may see lots and lots of fizzing bubbles coming from the threads. Is the screw changing color? Leave it in the liquid for a while and see what happens.

9 After about an hour, look at the pennies on the paper towels. What's happened to the ones you rinsed? What's happened to the others? What color is the paper towel under the unrinsed pennies?

What's Going On?

Why did the pennies look dirty before I put them in the vinegar?

Everything around you is made up of tiny particles called *atoms*. Some things are made up of just one kind of atom. The copper of a penny, for example, is made up of copper atoms. But sometimes atoms of different kinds join to make *molecules*. Copper atoms can combine with oxygen atoms from the air to make a molecule called *copper oxide*. The pennies looked dull and dirty because they were covered with copper oxide.

Why did the vinegar and salt clean the pennies?

Copper oxide dissolves in a mixture of weak acid and table salt—and vinegar is an acid. You could also clean your pennies with salt and lemon juice or orange juice, because those juices are acids, too.

Why did the unrinsed pennies turn blue-green?

When the vinegar and salt dissolve the copper-oxide layer, they make it easier for the copper atoms to join oxygen from the air and chlorine from the salt to make a blue-green compound called *malachite*.

How did the nail and the screw get coated with copper?

To understand how the nail and screw got coated with copper, you need to understand a little bit more about atoms. Atoms are made up of even smaller particles called *protons, neutrons,* and *electrons*. Electrons and protons are both electrically charged particles. Electrons are negatively charged and protons are positively charged. Negative charges attract positive charges, so electrons attract protons.

When you put your dirty pennies in the vinegar and salt, the copper oxide and some of the copper dissolve in the water. That means some copper atoms leave the penny and start floating around in the liquid. But when these copper atoms leave the penny, they leave some of their electrons behind. Rather than having

whole copper atoms in the liquid, you've got copper *ions*, copper atoms that are missing two electrons. These ions are positively charged.

Now add two steel nails and a screw to the mixture. Steel is a metal made by combining iron, other metals, and carbon. As you found out when you cleaned your pennies, your mixture of salt and vinegar is really good at dissolving metals and metal oxides. When you put the steel nail in the mixture, some of the iron dissolves. Like the copper atoms, each of the iron atoms that dissolves leaves two electrons behind. So you've got positively charged iron ions floating in your vinegar with the positively charged copper ions.

Originally, the steel nail was neutrally charged—but when the iron ions left their electrons behind, the nail then became negatively charged. And remember what we said way back at the beginning of this section: negative charges attract positive charges. The negative charges on the nail attract positive charges in the liquid. Both the iron ions and the copper ions are positively charged. The copper ions are more strongly attracted to the negative charge than the iron ions, so they stick to the negatively charged nail, forming a coating of copper on the steel.

Why did bubbles come off the steel screw?

Each water molecule is made up of two hydrogen atoms and an oxygen atom. In an acid (like vinegar or lemon juice), lots of hydrogen ions (hydrogen atoms that are missing an electron) are floating around. In the chemical reactions at the surface of the screw, some of these hydrogen ions join and form hydrogen gas. The bubbles that you see coming off the screw are made of hydrogen gas.

Mold Terrarium

Watch tiny blue, green, and white plants grow on leftover food.

What Do I Need?

- A clear container with a lid. (Big glass jars and clear plastic containers work great, but you'll have to throw away the container when you're through, so check with a grown-up about what you can use.)
- Adhesive tape
- Water
- Some leftover food (you can use whatever is in your refrigerator), such as bread, fruit (like oranges, lemons, or grapes), vegetables (like broccoli, zucchini, or green pepper), cheese, and cookies or cake

This Is Important!
DO NOT use anything with meat or fish in it—after a few days, these would start to smell very, very bad.

What Do I Do?

1 Ask a grown-up for 4 or 5 different pieces of leftover food. If the food is small—a grape or one section of an orange—use the whole thing. Cut bigger foods like bread or cheese into 1-inch chunks.

2 Dip each piece of food into some water and put it into your container. If you use a big jar, lay it on its side. Try to spread the pieces out so that they are close to each other, but not all in a heap.

3 Put the lid on the container. Tape around the edge of the lid to seal it.

4 Put the container in a place where no one will knock it over or throw it away. You may want to label it "Mold Terrarium."

Wow! I Didn't Know That!

When most foods get moldy, it means they aren't good to eat any more. But some cheeses are eaten only after they become moldy! Blue cheese gets its flavor from the veins of blue-green mold in it. When a blue cheese is formed into a wheel, holes are poked through it with thin skewers. Air gets into these holes, and a very special kind of mold grows there as the cheese ripens.

5 Every day, look at the food in your Mold Terrarium. For the first 2 or 3 days, you probably won't see much. But soon you should see blue or green or white fuzzy stuff growing on some of the pieces of food.

Here are some things to notice in your mold terrarium:

- What food started getting moldy first?
- What color is the mold? How many different colors do you see?
- What texture is the mold—flat, fuzzy, bumpy?
- Does everything in your Mold Terrarium get moldy?
- Does mold spread from one piece of food to another?
- Do different kinds of mold grow on different types of food?

MOLD TERRARIUM — DO NOT OPEN!!!

6 After a few more days, some of the food in your mold terrarium may start to rot and look really gross. You can watch how the mold spreads and how things rot for about 2 weeks. After that, it'll get boring, because not much more will happen.

What's Going On?

What is mold, anyway?

That fuzzy stuff growing on the food in your mold terrarium is mold, a kind of fungus. Mushrooms are one kind of fungus; molds are another.

Unlike plants, molds don't grow from seeds. They grow from tiny spores that float around in the air. When some of these spores fall onto a piece of damp food, they grow into mold.

Green plants are green because they contain a chemical compound called *chlorophyll*. Chlorophyll makes it possible for green plants to capture the energy of sunlight and use it to make food (sugars and starches) from air and water. Unlike green plants, mold and other fungi have no chlorophyll and can't make their own food. The mold that grows in your mold terrarium feeds on the bread, cheese, and other foods. The mold feeds itself by producing chemicals that make the food break down and start to rot. As the bread rots, the mold grows.

Ick! Who wants this stuff around?

It can be annoying to find moldy food in your refrigerator. But in nature, mold is a very useful thing. Mold helps food rot, which is an icky but necessary thing. In a natural environment, rotting things return to the soil, providing nutrients for other plants. Mold is a natural recycler.

Why does the mold on different foods look different?

There are thousands of different kinds of molds. One mold that grows on lemons looks like a blue-green powder. A mold that grows on strawberries is a grayish-white fuzz. A common mold that grows on bread looks like white cottony fuzz at first. If you watch that mold for a few days, it will turn black. The tiny black dots are its spores, which can grow to produce more mold.

Why didn't some foods get moldy?

If you used foods that contain preservatives, mold may not have grown very well on them. If you want to experiment more with mold, you can make one mold terrarium using food with preservatives (like a packaged cupcake) and another using food that doesn't have preservatives (like a slice of homemade cake). Which one grows more mold? You can also experiment with natural preservatives like vinegar and salt. If you do more experimenting, let us know what you discover!

Looking for Changes

The world is always changing. Many changes take a long time—but if you stop and look, you can catch the world in the act of changing. It doesn't matter where you live; you can find evidence of change just about anywhere you look.

The Science-at-Home Team Looks for Changes

The Science-at-Home Team went looking for change near the Exploratorium. First, we walked around the lagoon right beside the museum. On the cement columns and the trees, Linda found moss growing in moist, sheltered places where it was protected from

sun and wind. In cracks in the cement, she noticed tiny green plants growing. On a few trees, Ellen saw shelf fungi growing from the bark.

The fungi and the moss were helping break down the surfaces on which they grew. The roots of the tiny plants were slowly pushing against the cement, making the cracks imperceptibly bigger. We didn't wait around to see the results of all this activity—it takes years for a fallen log to decay on the forest floor, and we figured that moss breaking down cement would take a lot longer. So we headed down the residential street that leads toward the lagoon.

Ellen started looking for rust—and once she started looking, she found it everywhere. The best places to look, she discovered, are

edges and boundaries, like the places where two metal surfaces join, or where there is a crack or a hole in a piece of metal that makes an edge in the middle of a smooth surface. We all started looking for rust then—peering closely at metal trash cans, gratings, and cars parked on the street. The paint on the cars protects the metal from rust, but wherever the paint was chipped or scratched, we could see the telltale reddish-brown color of rust.

Linda explained that iron rusts when the iron atoms combine with oxygen from the atmosphere to make iron oxide, a scientific name for rust. Steel is an alloy, a metal made by combining iron, other metals, and carbon. When the iron atoms in steel combine with oxygen, the steel rusts.

While we were studying cars and gratings, a woman walked by and asked what we were doing. When we told her, she pointed out some wonderful rust streaks on a picket fence down the street. Each board of the fence was held in place by a nail. Rain had rusted the nails and left long streaks of rust running down the wood.

What you find in your neighborhood will depend on where you are. People who live in the city may find grass pushing its way through cracks in the pavement—but people in the country are more likely to find fungi growing out of decomposing logs.

Here are some of the changes our Home Scientists found in their neighborhoods:
• Mildew on books left in a damp basement
• A hole in a sidewalk made by water dripping from an air conditioner
• Soap turning into bubbles
• Veins of quartz pushing through cracks in rocks
• Tar melting in the sun
• Labels fading on bottles in the sun
• Worms and insects changing scraps into compost
• Polliwogs changing into frogs
• Cake-making ingredients turning into cake!

Take a look around you and see what changes you can find!

Made in the Shade

If you've ever watched your shadow on a sunny day, you know that when you move, your shadow moves with you. You can't just leave it on the floor or on a wall and walk away.

Except at the Exploratorium.

Not far from the Science-at-Home Team's offices is an exhibit called the *Shadow Box*. It's a place where people can leave their shadows on the wall. The *Shadow Box* is a big room, open at the front, with three shiny, greenish-yellow walls. The stuff that covers the walls is sensitive to light. When a bright light shines on these walls, they absorb some of the light and turn bright green for a few minutes. Then they fade back to a dull greenish-yellow.

People stand inside the *Shadow Box,* lean against the walls, and wait. About once a minute a buzzer sounds, then a very, very bright light flashes for a fraction of a second.

After the flash, the walls of the *Shadow Box* are bright green—except in those places where the light-sensitive material was blocked by people's shadows. In those places, dark body shapes decorate the inside of the box.

Some people wait with their arms stretched out, or their legs bent, trying to make a really weird shadow. Others like to stand still until the buzzer sounds, then leap high into the air, trying to catch their shadows in mid-jump. Then everyone stands back, leaving their shadows hanging on the wall. They admire them until the shapes begin to fade. A few minutes later, when the walls are blank again, the next buzzer sounds. Then more people leap in the air and laugh at the strange shadows they've left on the wall.

You'll have to come to the Exploratorium to walk away from your shadow, but this chapter has some amazing shadow tricks you can try at home. You can play with the shadows cast by the sun; make a Shadow Lab to experiment with the shapes, sizes, and textures of shadows; and create a theater where shadow puppets are the actors. Then you and your family can go on a shadow hunt to find the best shadows in your neighborhood—even after the sun goes down.

Sun Shadows

Use your backyard to play a shadow game or create a piece of shadow art.

What Do I Need?

- A sunny day
- A yard, driveway, or playground
- A watch or clock
- Colored chalk (at least 3 different colors)
- String
- Crayons
- A big piece of paper (butcher paper or an unfolded paper grocery bag)
- An object with an interesting shape (a bottle, toy truck, or action figure)

What Do I Do?
Shifting Shadows

1 Find a place where the shadow of your house or garage makes a straight line. Mark that line with a piece of chalk. (Use a piece of string if the shadow falls on grass.) Look at a clock to see what time it is.

2 Guess where the shadow will be in 15 minutes. Mark your guess with another color of chalk. (If you use string, make a mark on the "guess string" so you can tell the two strings apart.)

3 In 15 minutes, use a third color of chalk (or another piece of string) to mark where the shadow *really* is. How close was your guess? Want to try again? You can do this for as long as there's a shadow.

4 If you want, you can make this a contest with your friends. Mark the shadow of the house, then have each person mark her guess with a different color of chalk. Wait 15 minutes, and see whose guess was closest.

Tips for Home Scientists
These activities won't work very well around noontime. At noon, when the sun is high in the sky, shadows will just look like dark puddles around the bottom of things. In the morning and in the afternoon, when the sun is lower in the sky, things will have much longer shadows.

Shadow Sketches

1 Find something with an interesting shape—a bottle, a doll, a toy truck, an action figure. Go outside and find a place where you can see your shadow. Put the big piece of paper on a driveway or a sidewalk, or on a picnic table, if there is one. Stand the thing you're going to draw in the middle of the paper.

2 Use a crayon or chalk to trace around the thing's shadow. Is the tracing the same size as the thing? Is it the same shape?

3 Wait 20 minutes, then trace the thing's shadow again, using a different color crayon or chalk. Has the shadow moved? Has its size or shape changed?

4 Keep tracing the shadow every 20 minutes until you like the picture that you've drawn on the paper. Hang your art on a wall or on the refrigerator.

5 If you want to keep experimenting with shadow art, try some of these changes:

- Trace the shadow every 10 minutes, or every 30 minutes, or once an hour, or once every 2 hours.
- Trace all the shadows with the same color, then use different colors to fill in the outlines.
- Put 2 things on the paper and trace both shadows. Do the shadows start to overlap?

What's Going On?

Why bother to watch shadows?

All day long, shadows are shrinking and growing and changing shape. Late in the day and early in the morning, shadows change size and shape much faster than you think—as you find out when you try the Shifting Shadows activity.

Most people don't pay much attention to shadows and how they change. That's because most people don't think shadows are important. But the movement of shadows in sunlight is evidence of something very important, something that it took people thousands of years to figure out—the earth spins like a top and orbits around the sun.

Every morning, the earth spins or rotates so that the side you are on faces the sun. As the earth spins, you see the sun come up over the eastern horizon. The earth keeps spinning, and the sun seems to move across the sky. When the earth turns so that the side you are on faces away from the sun, you see the sun disappear below the western horizon. From where you are, on the surface of the earth, it looks like the sun is moving. But it's really the movement of our own planet that makes the sun seem to move and the shadows change.

Making a Sun Clock

Before there were clocks, people used shadows to tell time.

What Do I Need?

- A sunny day
- A pencil
- A compass
- A photocopy of page 69 (Don't tear the page out of the book!)

What Do I Do?

1 Ask a grown-up to help you with this experiment. On a sunny day, go outside with a compass, pencil, and photocopy of page 69. Put your compass on the ground and turn it so that the arrow and the "N" (for "North") line up.

2 Follow the directions on page 69 to find out how to line up the Sun Clock with your compass. Once you have the Sun Clock pointed in the right direction, you can figure out what time it is.

Tips for Home Scientists

If none of the cities on the list are close to you, you can still use your Sun Clock. See page 70 to find out how.

Start Here!

1 Put a compass on the ground and notice which way the arrow points.

A	**B**	**C**
Washington, DC New York, NY	Atlanta, GA Miami, FL Chicago, IL Louisville, KY Columbus, OH Detroit, MI	Minneapolis, MN Kansas City, KS Dallas, TX New Orleans, LA
D Denver, CO	**E** San Francisco, CA Los Angeles, CA Salt Lake City, UT Rapid City, SD	**F** Seattle, WA Boise, ID

2 Find the city that's closest to where you live and see what column it's in.

3 Put this page on the ground and move it around until the arrow that represents your city points the same way as the arrow on your compass.

Geographic North

A B C DE F

Sun Clock

June 1 – July 1
May 1 – August 1
April 1 – September 1
March 1 – October 1
February 1 – November 1
January 1 – December 1

4 Stand a pencil on today's date and look at the shadow to find out what time it is.

Daylight Saving Time

Standard Time

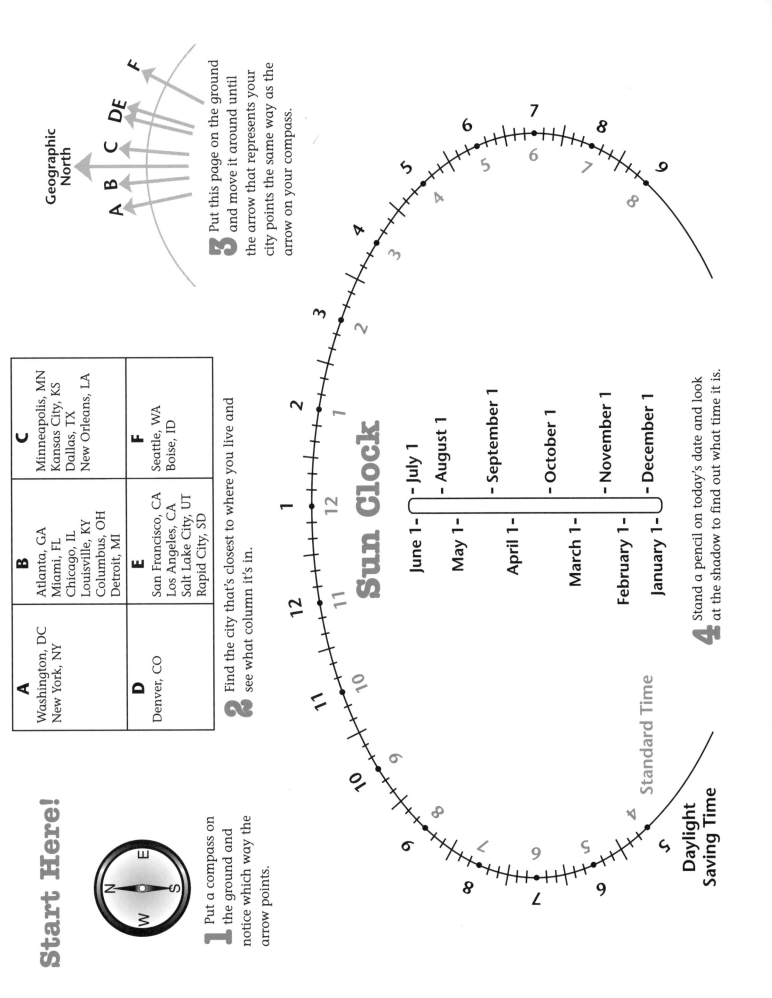

What's Going On?

Why does it matter what city I'm in?

What time it is depends upon where you are on the planet. That's why you use a compass to orient yourself in this activity.

A compass needle (which is attracted to the magnetic field of the earth) points in a direction called *magnetic north*. That isn't exactly the same as *true north*, or *geographic north*, which is the direction of the earth's North Pole. We've set up our Sun Clock so that it uses geographic north as a reference point. If you don't line up the page with geographic north, the Sun Clock won't give you the right time of day.

The difference between magnetic north and geographic north is called *magnetic declination*, and it's different in different locations. When you position the Sun Clock according to the directions on page 69, you are compensating for the magnetic declination of where you live. After you do this, the "Geographic North" arrow at the top of the page will be pointing to geographic north and your Sun Clock will work just fine.

What if my city is not on the list? Can I still use my Sun Clock?

If you aren't near any of the cities on the list, you can still use the Sun Clock. Go out at night and look for the North Star. (You may need a book of constellations to help you find it.) Mark an arrow on the ground that points toward the North Star. That's geographic north. The next day, position your photocopy of the Sun Clock (page 69) with the Geographic North arrow (in the top right-hand corner of the page) pointing in the same direction that you marked on the ground (toward geographic north). Now follow the rest of the instructions on the page.

How does a shadow tell time?

Shadows change direction, depending upon the time of day. A Sun Clock like this one uses a shadow's position to tell the time.

Why doesn't the time on my Sun Clock exactly match the time on my watch?

The time you get from your Sun Clock is solar time, not standard time. The two aren't exactly the same.

According to *solar time*, it's noon when the sun reaches its highest point in the sky. But the sun is always moving across the sky—which means that noon where you are is at a slightly different time than noon at a place a few miles to the east or west.

Back before 1883, people used solar time. Each community kept its own time, basing that time on the sun's position in the sky. Back then, noon in one town would be four minutes later than noon in a town fifty miles to the east.

In 1883, to regulate time for the sake of railroad schedules, the United States adopted what is called *standard time*, designating time zones and requiring all communities within a time zone to keep the same time—even though that standard time didn't quite match solar time.

If you are in the middle of your time zone, your Sun Clock will be fairly accurate. If you are at one edge of your time zone, the time on your Sun Clock (solar time) may differ from the time on your watch (standard time) by as much as forty minutes.

Why do I have to put the pencil on different spots for different times of the year?

The position and length of a shadow depends on the time of day—but it also depends on the season of the year. That's because the sun's position at a certain time of day is different in different seasons. If you want to know more about this, try Charting the Sun's Path on page 71.

Charting the Sun's Path

If you are very, very patient, you can use a shadow to see how the sun's path changes with the season.

What Do I Need?

To work outside:
- A flagpole or fence post in an open area of asphalt (like a playground or a driveway)
- Paint that won't wash off in the rain
- A reliable watch or clock
- Lots of patience—this experiment takes a year to complete!

To work inside:
- A south-facing window
- A small mirror
- Masking tape
- Some Post-it notes
- A reliable watch or clock
- Lots of patience—this experiment takes a year to complete!

What Do I Do?

1 Before you start, ask a grown-up to read this activity with you and figure out whether you should do it indoors or outdoors. You'll also need permission to make paint marks on the asphalt if you work outside, or put tape on the walls if you work inside.

2 Figure out a time of day when you will be at home at least once a week. The sun must be out at this time of day, so if you're starting in summer, don't choose a time near sunset. (The sun sets earlier in the winter.)

3 **Outside:** Find the shadow of something that doesn't move (like a flagpole or a fence post) in an open area of asphalt (like a playground or a driveway).

Inside: Find a south-facing window. Make sure the sun shines through your window at the time of day you've chosen.

4 **Outside:** At the time of day you've chosen, mark where the end of the shadow falls by making a spot of paint on the asphalt.

Inside: Use the masking tape to cover most of the mirror, leaving a small square open to reflect the sun. Put the mirror on the windowsill and position it so that it reflects a spot of sunlight onto the wall. (If you can, fasten the mirror to the windowsill so that no one will move it by accident. Make sure everyone in your family knows about your experiment!) Mark where the spot of sunlight falls. We used Post-it notes, and wrote the date on each one. You can use tape if you get a grown-up's permission.

5 **Outside:** A week later, do it again. The shadow will have moved a little bit.

Inside: A week later, do it again. The spot of light will have moved a little bit.

6 Do the same thing once a week for a whole year. As you are doing this, you have to take daylight saving time into account. If you move your clock forward to adjust for daylight saving time, change the time that you make your weekly mark by one hour.

7 At the end of the year, take a look at all the marks. If you connected them, what shape would they make?

Every day, the sun takes a slightly different path through the sky. On the longest day of the year (the "summer solstice"), it arches up high. On the shortest day of the year (the "winter solstice"), it stays down low.

What's Going On?

What shape do your marks make?

At the end of the year, connecting your marks will make a figure-eight pattern like the one on the right. This pattern is called the *analemma*, which is Latin for "sundial." The analemma is a visual record of the sun's changing position in the sky over the course of the year.

Why isn't the sun in the same place at the same time every day?

People usually say that the sun rises in the east and sets in the west. That's not exactly true. If you were to watch the sun rise each morning over the course of an entire year, you'd see that the sun doesn't always rise in the same place. In the summer, the sun rises a little bit north of due east. The date on which it rises the farthest to the north is June 21, the summer solstice and the longest day of the year. In the winter, the sun rises a little bit south of due east. The date on which it rises the farthest to the south is December 21, the winter solstice, the shortest day of the year.

Suppose you watched the sky on the day of the summer solstice and the winter solstice. The path of the sun across the sky on these two days would look something like the picture on page 72. On the summer solstice, the sun rises much higher above the horizon, taking a longer path across the sky. On the winter solstice, the sun never gets as high in the sky.

The sun's path across the sky changes with the seasons partly because the earth's axis—the imaginary line through the earth around which our planet spins—is tilted with respect to the earth's orbit around the sun. Take a look at the picture below. It shows the positions

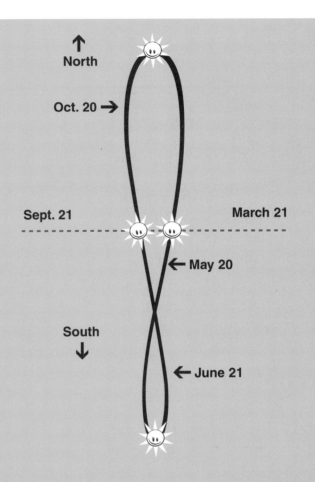

of the earth and sun on the summer solstice and the winter solstice. When it's summer in the Northern Hemisphere, the North Pole is tilted toward the sun and the sun shines on a greater area of the Northern Hemisphere. As the earth spins, places in the Northern Hemisphere stay in the sunlit area longer, and that means the days are longer. When it's winter in the Northern Hemisphere, the North Pole is tilted away from the sun, the sun shines on a smaller area of the Northern Hemisphere, and the days are shorter.

The earth's tilt affects the position of the sun in the sky—and so does the shape of the earth's orbit around the sun. The earth's orbit is *elliptical* (oval), with the sun near one of the ends of the ellipse. This means that the earth is sometimes closer to the sun and sometimes a couple of million miles farther away. When the earth is closer to the sun, it moves a little bit faster in its orbit. When the earth is farther away, it slows down, just a bit. This variation in speed affects when the sun is at a particular position in the sky—and that affects where the shadows fall.

Shadow Lab

Experiment with the shapes and sizes of shadows.

What Do I Need?

- Two flashlights (penlights or Mini-MagLites work best, but an ordinary flashlight is okay)
- A white sheet or tablecloth
- A room you can make dark
- Notebook paper
- Scissors
- A small box (like a box of crayons)
- A glass measuring cup with painted lines
- A wineglass

What Do I Do?

1 Put the sheet down on a table. Stand the box in the middle of the sheet. Turn on a penlight. Turn off all the other lights in the room.

2 Hold the penlight down on the table, about a foot away from the box. Shine the light on the box, and slowly move the light up and over it. Watch the shadow. Does the shadow stay the same shape as the box? Can you make the shadow bigger and smaller? Can you make the edges of the shadow sharp or fuzzy? What happens if you move the light in a circle around the box?

3 Shine the light on one side of the box. Now turn on the second light, and shine it on the other side of the box. How many shadows can you make? Can you make the shadows overlap? Can you make one shadow darker than the other?

4 Put the measuring cup on the table. Shine the light through it from the side. Can you read the shadow words and numbers on the shadow cup? Hold the light right over the cup, then move it down until it's inside the cup. What happens to the shadow?

The Harkins family made their own "planetarium," shining the light up through colanders and sieves and onto the ceiling. The Stevenson family said their favorite shadow-makers were their hands!

5 Shine the light through the wineglass from the outside and inside. Now fill the glass with water and shine the light through it. How does that change the shadow?

6 Make a snowflake from a square of paper. Fold the square in half, then fold it in half again. Fold the little square into a triangle, then fold that triangle again, if you can. Cut little tiny snips out of the folded edges. Unfold it to make a snowflake. Place the snowflake over the wineglass and shine the light down through it. Move the light and watch how the shadows change.

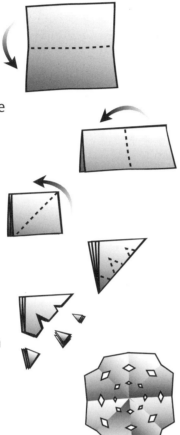

7 Loosely crumple up a piece of paper and put it down on the table. Move the light and watch how the shadows on the paper change.

8 Some other fun things to play with on your shadow table are combs, salt shakers, wax paper, facial tissue, glass bowls, and kitchen utensils like forks, colanders, and spatulas.

What's Going On?

Why should I use a penlight rather than a regular flashlight?

To make sharp, clear shadows, you need a small, bright light source. A scientist would call such a light a *point source*, since all the light comes from a single point. A penlight is a point source. Most regular flashlights have reflectors that make a broad beam of light—which is nice if you want to find your way in the dark, but not as good if you want to make sharp-edged shadows.

Why does filling the wineglass with water change the shadow?

When the wineglass is full of water, it acts like the lens in a magnifying glass. It bends all the light shining through the wineglass so that the light all lands on one spot, making that spot very bright.

Shadow Theater

Create a stage of shades—and make your own shadow show

What Do I Need?

- A shoe box (or other small, sturdy cardboard box)
- Masking tape
- A white plastic grocery bag
- Two penlights (regular flashlights will work, but not as well)
- A chair
- A room you can make dark
- A paperback book or small box
- Some interesting objects (see the list on page 78 for some ideas)
- Construction paper or file cards
- Scissors
- Paper clips

What Do I Do?

Making the Theater

1 Have a grown-up cut a rectangular hole in the bottom of the shoe box. Leave about 1 inch on all sides.

2 To make the screen for your theater, cut the handles off the white plastic grocery bag. Cut down the middle of both sides, and then across the bottom of the bag. You'll end up with two big pieces of white plastic. Use the side of the bag that DOES NOT have lettering.

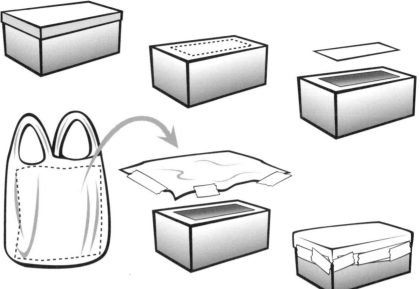

3 With the cut-out bottom of the box facing up, tape one edge of the plastic to one side of the box. Pull the plastic tightly over the hole, and tape the opposite edge to the other side of the box. (Don't pull it so tightly that you bend the box!)

4 Tape the loose flaps of plastic to the ends of the box. (Fold or cut away extra plastic if you need to.) Tape all the way around the box to make sure the plastic is on tight and won't come off. Now you've got a screen where your audience can watch shadows.

5 Put the box on a table so that the screen faces the audience. Tape the penlight to the top of a chair about 4 feet behind the table. Move the chair so that the light shines into the back of the box and onto the screen. Tape the clip on the penlight down tight so it will stay in the ON position.

Wow! I Didn't Know That!

For more than 1,000 years, people on the Indonesian islands of Java and Bali have enjoyed *wayang*—shadow puppet plays. The puppets are made of pieces of stiff leather attached to thin sticks. Holes are cut in the leather to create facial features. The arms and legs are jointed so that they'll bend and move. A bright light behind the puppets projects their images onto a screen of thin white cloth.

Creating Your Own Special Effects

1 Turn the room lights off and experiment with the shadows you can make on the screen. You can make some great special effects by shining the light through stuff from your kitchen.

2 Lay a paperback book or a small box down on the inside of your theater to be the "stage" that you put the things on. With a second penlight, you can make double shadows, shadows that move, and even stranger effects!

3 Ask a friend to be your audience and tell you how the effects look from the other side of the screen.

Here are some things we've tried:

- Clear glass jar (empty or filled with water)
- Glass jar filled with water and food coloring (green and blue make good underwater scenes; red is good for a Martian scene or to look like something's on fire)
- Glass salt shaker or crystal (can you make a rainbow inside a shadow?)
- A magnifying glass or a pair of eyeglasses used for reading
- A plastic deli container colored with permanent markers
- A colander
- A fork or spatula
- Crumpled plastic wrap

When they made their Shadow Theater, Home Scientists in the Turner family experimented with different kinds of light, including candlelight. The Stevenson family made a giant Shadow Theater! They hung up a big white sheet and used their bodies to make life-sized shadow plays.

Putting on a Play

1 First, figure out a story you want to tell. You can make a shadow play of a story you already know—like a fairy tale or your favorite book. Or you can make up your own story. Try to think of a story that will use the special effects you liked best.

2 To make puppets for your play, cut any shape you want—a person, a bear, a space monster, a truck—out of construction paper or a file card. Make sure your puppets are the right size to fit inside your Shadow Theater. Unbend a paper clip to make a long, thin wire. Tape one end of the wire to the back of the shape. Hold the other end of the wire in your fingers.

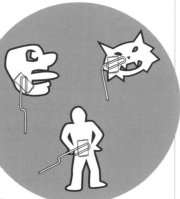

Hold your puppet between the light and the screen. The wire can come out of the top, bottom, or sides of the puppet. Experiment and see what works best.

3 Move the puppet to make its shadow grow and shrink on the screen. If you have a second light, you can make one puppet have two shadows. Remember, your audience can't see the puppets—just their shadows. With a little practice, you can trick your friends into seeing things that aren't really there!

4 If you want scenery for your play, cut shapes that look like buildings or trees out of construction paper or file cards. Tape them to the bottom of the theater. If you tape them close to the screen, they'll make sharp shadows. If you tape them farther back, they'll make fuzzier shadows. Experiment with what looks best.

5 Use your puppets, scenery, and special effects to put on a shadow show that will amaze your family and friends!

What's Going On?

Why make a Shadow Theater?

If you make a Shadow Theater, you can experiment with shadows and then use those shadows to put on a show. At the Exploratorium, we like to do things that combine art with science. Both artists and scientists spend their time noticing interesting things about the world. By making a Shadow Theater, you can be both an artist and a scientist—and have a good time, too!

Walking in the Shadows

Now that you've experimented with making shadows, you may want to go for a walk and see what different shadows you can find in your neighborhood.

The Science-at-Home Team Watches Daytime Shadows

After work one day, the Science-at-Home Team took a walk through the neighborhood around the Exploratorium. Because it was late in the afternoon, and the sun was low in the sky, our shadows were long and narrow. Pat found out that if she stood at the top of one steep hill, her shadow was more than half a block long!

We soon found out that we could make our shadow bodies do things our solid bodies could never do. Ellen liked to make her shadow bend around the corners of walls and doorways, or zigzag up stone stairs and curbs.

Jason, who drew the pictures for this book, told us about a cool shadow trick he used to do as a kid—the Giant Eyeballs. Ellen and Pat stood right next to each other. They each put their hands on top of their heads, elbows sticking out, and scrunched their necks down into their shoulders. On the white wall of a garage, their shadows really did look like giant eyeballs! When they tilted their heads toward each other, the giant eyes crossed. When they raised their hands, the eyes looked surprised.

After that we stopped for a rest under a shady tree. We noticed that we could see darker shadows where lots of leaves over-lapped. Between the leaf shadows, we saw bright circles of light. Linda explained that light shining through a tiny hole makes an image or picture of the light source. The circles of light we saw in the tree's shadow were actually images of the sun!

Linda also told us that if we looked at our hand shadows during a solar eclipse, when

the moon was covering part of the sun, each of those bright circles would look like a cookie with a bite taken out of it.

In front of one house, Pat noticed something very odd going on with her shadow. She was standing near a house where a tree cast a shadow on the sidewalk. The sun was in the east—and all the shadows we had seen slanted east. But in a patch of light within the tree's shadow, Pat was casting a shadow that slanted west! How could that happen?

It took us a few minutes to figure out what was going on. Sunlight was reflecting off the window of the house, above Pat's head, and making a patch of light in the tree's shadow. Pat was blocking some of that light to make a west-slanting shadow in the reflected patch of light.

There were other patches of light in the tree shadow, where sunlight was shining between the leaves and branches. Ellen discovered that she could block some of this light and make an east-slanting shadow right beside Pat's west-slanting shadow.

We played with the weird double shadows until we got hungry. That seemed like a good time to take a break and have dinner. We'd continue our shadow hunt after dark.

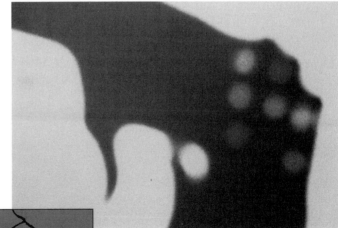

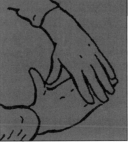

If you let sunlight shine through the holes between your crossed fingers, you'll see lots of little images of the sun. That's what those blurry round blobs of light are in the shadow of your hand.

The Science-at-Home Team Looks for Shadows at Night

By the time we'd finished eating, it was dark outside—or as dark as it ever gets in the city. We stood out on the sidewalk for a minute and pointed to all the different kinds of lights and shadows we could see. Streetlights, car lights, and light shining through the windows of stores and restaurants filled the sidewalk with a wonderful array of constantly changing shadows.

The streetlights were spaced about thirty feet apart. Ellen noticed that when she stood beneath a streetlight, she had one very solid-looking shadow pointing away from the light. When she walked toward another streetlight, she had two less distinct shadows pointing in opposite directions. As she neared the second light, the first shadow disappeared and the second one got darker.

Pat stood between two streetlights for a few minutes and watched as her two shadows were joined by a third shadow—and sometimes a fourth—as the lights from passing cars washed over her. Linda found a corner where she had three shadows, all going in different directions, cast by three lights—two on one street and one across the intersection.

As we walked along, we talked about how different shadows look at night and how they can sometimes seem a little scary. A lot of the time, a shadow may not look anything like the real object. Ellen watched as a car went by and made her shadow huge. She's not very tall, but her shadow looked like a giant's.

Our shadows were beside us, behind us, or in front of us, depending on what lights we were walking by, or what lights were driving by us. The most fun, we thought, was when a car drove by. First we were walking ahead of our shadows, then all of a sudden they got very long, raced by, and were ahead of us. Then they were gone. When we got back to the Exploratorium, we agreed that the shadow walk had been one of our favorite field trips. Jenefer, a teacher who worked with the Science-at-Home Team, said that she planned to take her kids on a long daytime shadow walk. Since she lives out in the country, where there aren't any streetlights, she said they'd have to try some nighttime shadow play in the backyard.

What shadows can you and your family find around your neighborhood?

What Do You Say?

"The sorry mountain understands simple pity."
"Proud candle simply discovers soft darkness."
"Joy grows against harsh stormy wonder."

These three groups of words may not make a lot of sense to you, but they still sound like sentences. They're from an Exploratorium exhibit called *Reaching for Meaning*. At this exhibit, visitors reach into a large drum filled with long wooden blocks, pull out half a dozen blocks, and arrange them on a table. Words are painted on the 4 long sides of each block. Blocks with words painted in red are nouns. Blocks with yellow words are verbs. Blocks with green words are adjectives.

Visitors use these blocks to build sentences. As long as the blocks are arranged in certain patterns—like adjective, noun, verb, adjective, noun—they'll make a sentence, no matter what the words on the blocks mean.

Scientists take things apart to see how they work. A scientist who takes apart a machine finds parts like cams and gears. A botanist who dissects a vegetable finds parts like seeds and roots. Linguists, scientists who study language, take apart sentences and find parts of speech like nouns and verbs. By looking carefully at how a sentence is put together, linguists discover the structure of a language. They use this information to explore connections between different languages, and to learn how people name things, give information, and communicate ideas.

Many of the activities in this chapter are home versions of Exploratorium language exhibits. With them you can have fun exploring some of the ways that you use language every day. You can experiment with tongue twisters, communicate with pictures instead of words, and invent your own alphabet.

Mouth Moves

Here are some fun ways to trick your tongue.

What Do I Need?

- Your mouth
- The list of tongue twisters on the next page

What Do I Do?

1 Say the alphabet, out loud, slowly. Feel how your tongue and your lips move as you say each letter.

2 Keep your tongue up against the roof of your mouth and say the alphabet again. Does it sound strange? What letters are hard to say? Are there any letters you can't say at all? Can you say your name?

3 Keep your tongue down at the bottom of your mouth and say the alphabet again. Now what letters are hard to say? Can you say your name?

4 Keep your mouth open. It's okay to move your tongue, but don't shut your lips. Can you say the whole alphabet that way? What letters are hard to say? What letters are impossible? Can you say your name?

Wow! I Didn't Know That!

Ventriloquists are people who have learned to speak clearly without moving their lips or jaws. They move their tongues inside their mouths to produce all the sounds. Ventriloquists do not really "throw their voices." Because her mouth doesn't move, when a ventriloquist like Shari Lewis talks, it just seems as if the voice is coming out of her puppet's mouth.

A B C D E F G
H I J K L M N O P
Q R S T U V
W X Y Z

5 When you talk, your mouth and your lips and your tongue all work together to make one sound right after another, so you can say words and sentences. "Tongue twisters" are groups of words with sounds that are hard to say together.

Try saying these tongue twisters out loud. Some may be easy for you to say, but hard for someone else in your family. Others may be tricky for everyone.

These may look simple, but can you say each of them 5 times as fast as you can?

- **Peggy Babcock**
- **Red leather, yellow leather**
- **Mixed biscuits**
- **Selfish shellfish**
- **Greek grapes**
- **Toy boat**

Can you say each of these quickly? More than once?

- **A bloke's bike's back brake block broke.**
- **Shave a cedar shingle thin.**
- **A proper pot of coffee in a proper copper coffeepot**
- **Pass the plain pliable painted paper plate, please.**
- **"Mrs. Smith's Fish Sauce Shop," said the shiny sign.**
- **Which is the witch that wished the wicked wish?**

According to the *Guinness Book of World Records,* this is the hardest tongue twister in the English language. Can you say it?

- **The sixth sheik's sixth sheep's sick.**

Does the wristwatch shop shut soon?

Wow! I Didn't Know That!
English isn't the only language where people play with tongue twisters. The Schreyer family sent us these two international tongue twisters for Home Scientists to try.

This one is in German:
"Kaplane kleben keine Papp-Kaplan-Plakate."
(Preachers glue no cardboard-preacher-posters.)

This one is in French:
"Dina, dina dit-on du dos dodu d'un didon."
(Dina ate the meaty back of a turkey.)

Make Your Own Tongue Twisters

1 Write down a list of nouns, verbs, and adjectives that all start with the same letter. Then try to make some sentences using those words. The letters S, F, B, and P are good letters to try first.

For harder tongue twisters, make a list of words that have the same sounds in the beginning or middle or end of the word— like <u>b</u>ox, ro<u>bb</u>er, glo<u>b</u>. The sounds of *K, S, Sh, P, B,* and *F* are good ones to start with.

big black rubber baby buggy bumpers

2 You can play a tongue twister game if you want. You and a friend each make up some tongue twisters and write them down. Try to read one of your friend's tongue twisters 5 times. If you can, you get a point. If you can't, your friend gets a point. Then it's your friend's turn to try one of your tongue twisters.

What's Going On?

Why does keeping my tongue from moving change the sounds I make?

When you speak, air from your lungs pushes against your vocal cords, two bundles of ligament and muscle in your *larynx,* or voice box. The vocal cords vibrate and make a sound. (If you put your hand on your throat when you speak, you can feel those vibrations.)

The sound produced by the vocal cords echoes around in the cavities of your throat, mouth, and nose. By changing the position of your lips, jaw, tongue, and larynx, you change the shape of those cavities—and that changes the sound.

When you are first learning to speak, you learn to move your lips, jaw, tongue, and larynx to make all the sounds you need. Now you make those sounds into words without thinking about how you are shaping them. To become aware of what you are doing with your mouth when you talk, you have to stop doing it— stop moving your tongue or your lips. Suddenly, you are aware of movements that you usually ignore.

eee...

ooo...

ahh...

What's Going On?

What makes tongue twisters so hard to say?

Pronouncing a word is not just a matter of putting a collection of sounds together. It also involves modifying sounds so that they are easier to say together.

When you speak, your mouth and tongue and lips are forming one sound—and getting ready for the next sound that's coming up. For instance, the sound of the "e" in the word *bet* is similar to the sound of the "ea" in the word *bear*. Now say both words out loud, and pay really close attention to what your tongue is doing. When you say "bet," your tongue is getting ready to pronounce the "t," which changes the sound of the "e" slightly.

Tongue twisters trick your tongue into anticipating the wrong sound. Try saying *sheik* and pay attention to where your lips are positioned when you are saying the "ee" sound. Now say *sheep* and notice where your lips are. If your mouth is positioned to make the "k" sound, it's tough to make the shift to a "p" instead.

Tschaa-schtscht-pshmie-pftschie-nie!*

This blending of sounds to accommodate each other is one of the reasons that speech synthesizers (like the ones in computers that can talk to you) sound artificial. When the computer speaks, the sounds in a word are not modified by the surrounding sounds—and the resulting speech sounds odd.

The Polish tongue twister used during World War II was "Chrazaszcz brzmi w trzcinie," which sounds something like "Chahn-shcht bzhmee ftschee-neeieh" and means "The beetle buzzes in the thicket." Our German spy is attempting to say the tongue twister, but he's not pronouncing it right.

What good are tongue twisters, anyway?

By giving you a difficult series of sounds, tongue twisters make you aware of how you move your mouth to make sounds. Speech therapists have used tongue twisters to improve people's pronunciation, and dentists have used them to check the fit of dentures. Some people recommend them as a hiccup cure. A person suffering from hiccups is told to repeat one medium-length tongue twister five times without taking a breath.

During World War II, a tongue twister was used to catch spies. After Germany conquered Poland, the Polish Resistance used a tongue twister as its password. Any German spy who did not grow up speaking Polish would have great difficulty pronouncing the tongue twister.

I M O K. U R O K.

You can make sentences without words!

RUOK?

IMOK.

What Do I Need?

• The word puzzles on these two pages

What Do I Do?

1 Can you read this sentence?
    I C U.
I C U isn't a sentence. I C U isn't even a word. But if you say each letter out loud, you'll say a simple sentence, "I see you." Some letters in English sound just like other words. When you say C it sounds just like the words "see" and "sea."

2 Try to figure out what these sentences mean by saying each letter or number out loud. The answers are on the next page.

    A B S E-Z 2 C.
    K-T S B-Z.
    I M N A C-T.

Can you make up your own sentences using only letters and numbers?

3 A rebus is a kind of puzzle that adds pictures to letters and numbers to make words and sentences. As Arnold Schwarzenegger once said:

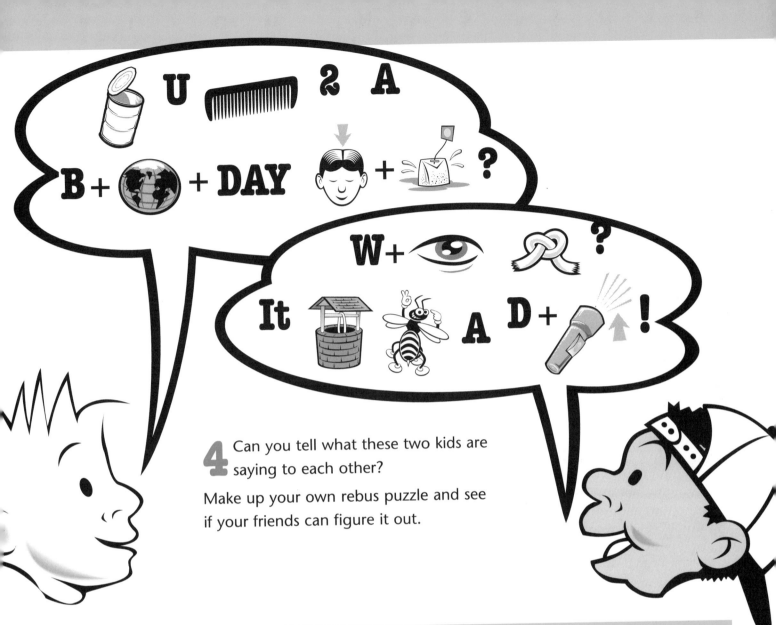

4 Can you tell what these two kids are saying to each other?

Make up your own rebus puzzle and see if your friends can figure it out.

What's Going On?

Making rebuses was an important step in the history of writing. Back in the fourth millenium B.C., in a region called Mesopotamia, the ancient Sumerians developed a system of writing with pictographs—they drew pictures to represent things that they wanted to keep records of.

But drawing pictures to represent things is a limited way of writing. You can write only about things that you can draw—like an ox or a bird or a fish or a plow. You can't write someone's name or write about something that you can't draw—feelings like love or hate or confusion.

To get around this limitation, the Sumerians started using pictures to make readers think of a sound, rather than an object. For instance, you can think of

 this picture as an eye, or you can just think of the sound that you make when you say the letter I. In Sumerian writing, a picture of a hand, for example, came to stand for the Sumerian word *su,* which meant "hand," but also stood for the sound "su," which could be used to make other words. Using this rebus writing made it possible to write about things that couldn't easily be pictured.

Answers
1. I see you.
2. A bee is easy to see.
   Katie is busy.
   I am in a city.
3. I'll be back.
4. Can you come to a birthday party?
   Why not? It will be a delight!

Writing Without Words

You can say a lot with pictures and symbols.

What Do I Need?

- The pictures on these two pages
- Paper to draw on
- A pencil, pen, or crayon

What Do I Do?

A Is for Alphabet

1 Very early systems of writing used a single picture or symbol to stand for a word. But that meant that people had to know a lot of different symbols if they wanted to send a long message. Almost 4,000 years ago, the Phoenicians came up with the idea of an alphabet. In an alphabet, each symbol stands for a single sound.

The Phoenician word for "ox" was *aleph*. The first symbol in their alphabet stood for the beginning sound in the word *aleph*. The symbol looked like the head and horns of an ox. Today a version of that same symbol is what we call the letter A—the first letter in the English alphabet.

2 You can make up your own personal alphabet. Think of some simple symbols that start with the letters A through Z. You might draw a moon (or a man or a monkey) to stand for the sound of M. You might draw a duck (or a dog or a dinosaur) to stand for the sound of D and an alligator (or an apple or an ax) to stand for the sound of A.

The name "Adam" might look like this:

Once you figure out all the symbols you need, you can share your personal alphabet with your friends and send messages to each other.

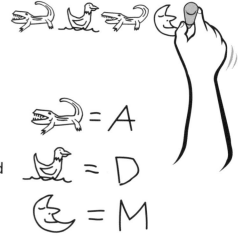

Signs and Symbols

1 Sometimes symbols don't stand for sounds at all. If you saw this sign, you probably wouldn't think of the letter B or the word "be." You'd think the sign was a warning that there might be stinging insects nearby.

2 Symbols that label something without spelling the word out with letters are called *pictograms.* You can see modern pictograms—sometimes called *international symbols*—in restaurants, on street signs, in cars, and on computers. See if you can match these symbols with what they label:

Printer

Girls' bathroom

No parking

Elevator

Heater

3 You can make up your own pictograms for things that happen around your house or neighborhood. Here are some that the Science-at-Home Team made up. See if you can match each pictogram with its message:

1. Warning: messy desk

2. I walked to work in the rain

3. Kids' Bubble Zone

4. No dogs on the couch

5. I had chocolate ice cream

What's Going On?

What does all this have to do with the alphabet we use?

It's easy to take the alphabet for granted, but it's really a remarkable invention. Somehow, this bunch of odd-looking squiggles communicates thoughts and ideas.

The alphabet we use today evolved from one developed in ancient Phoenicia, a region that is now part of Lebanon, Syria, and Israel. It had twenty-two letters, each one a picture of a common object. The letter represented the first sound in the word for that object.

The Greeks learned the alphabet from the Phoenicians sometime before 700 B.C. Because the Phoenician letters didn't include all the sounds that are used in the Greek language, the Greeks had to add a few letters of their own. The Greek alphabet was adopted by the Etruscans, a civilization that preceded the Romans in the region now known as Italy. In 509 B.C., the Romans conquered the Etruscans and later adopted the alphabet. Our alphabet is derived from the Roman version of that alphabet.

Silly Sentences

Make strange sentences by mixing and matching different kinds of words.

What Do I Need?

- At least 6 sheets of construction paper: one each of yellow, orange, red, green, blue, and brown (Don't use black.)
- A black marker, pen, or crayon
- Scissors
- A box or a hat or a big pot

What Do I Do?

1 Cut each sheet of paper into 8 strips about the same size.

2 You're going to write a different kind of word on each different color of paper. Here are the 6 kinds of words you'll need:

- **Yellow—Noun:** The name of a person, place, or thing—like *snake, football, bathtub, New Jersey.*
- **Red—Verb:** An action word—like *runs, spits, explodes, remembers.*

- **Green—Adjective:** A word that describes something or someone—like *messy, pretty, enormous, green*.
- **Orange—Adverb:** A word that tells how an action is done—like *carefully, slowly, happily, loudly*.
- **Brown—Conjunction:** A word that joins other words together—like *and, but, if, or, because*.
- **Blue—Preposition:** A joining word that tells where or when something is done—like *on, in, behind, about, at, without, for*.

If you want, a grown-up can help you make your word strips.

Tips for Home Scientists

When you are writing down your nouns and verbs, we suggest that you choose either singular nouns and verbs that agree with them *(girl runs)* or plural nouns and verbs that agree with them *(girls run)*. Don't mix the two.

3 On the front of each strip of yellow paper write one noun. Write a different noun on the back of each strip. You'll end up with 16 different nouns, all on yellow strips.

4 Do the same thing for the other 5 colors of paper and the other 5 kinds of words. You'll end up with 16 of each kind of word, each kind on its own color of paper. (If you can't think of 16 different words, just write as many as you can, or write some of them 2 or 3 times.)

5 Put all the strips of paper into a box. Shake the box to mix up the strips.

6 Reach into the box, and pick out 8 strips of paper. Put them down on the table in a line. Read the Silly Sentence out loud. Does it really make a sentence? Can you move the strips around so they do make a sentence?

You may not be able to use all 8 word strips you picked. If you get too many of one kind of word, pick 1 or 2 more strips to make a sentence.

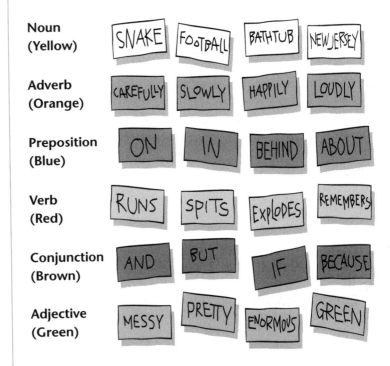

Noun (Yellow)	SNAKE	FOOTBALL	BATHTUB	NEW JERSEY
Adverb (Orange)	CAREFULLY	SLOWLY	HAPPILY	LOUDLY
Preposition (Blue)	ON	IN	BEHIND	ABOUT
Verb (Red)	RUNS	SPITS	EXPLODES	REMEMBERS
Conjunction (Brown)	AND	BUT	IF	BECAUSE
Adjective (Green)	MESSY	PRETTY	ENORMOUS	GREEN

7 Look into the box and pull out 8 strips in these colors, in order:

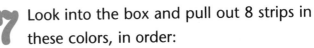

GREEN YELLOW RED ORANGE BLUE GREEN GREEN YELLOW.

Put the strips on the table in the same order.

8 Read that Silly Sentence out loud. Does it sound like a sentence? (It probably does, but it may be a very strange one.) What happens if you turn the strips of paper over? Do the new words still make a sentence?

Can you draw a picture to illustrate your favorite Silly Sentence?

Here's a Silly Sentence the Science-at-Home Team made:

"Friendly eggshell turns suddenly on pretty green swingset."

When we turned the strips over, we got:

"Purple spoon hides wildly beside enormous wooden potato."

What's Going On?

What do Silly Sentences have to do with science?

The science of linguistics studies language in all its facets, including how it is structured. Word games like Silly Sentences reveal a little bit about the structure of the English language.

Even if you don't think you know much about grammar, you probably know that this is not a sentence:

"Fast run girls little."

But this is:

"Little girls run fast."

You may not be able to state grammatical rules, but you use them to shape sentences when you talk or write. Most of the time, you know these rules without thinking about them. In fact, you can recognize a collection of words as a sentence, even if the idea that the sentence communicates doesn't make a lot of sense (like the friendly eggshell on the swingset above).

What else can I do when I'm playing with Silly Sentences?

Once you make a sentence using the words on one side of the paper strips, chances are good that the words on the other side will make a sentence, too. The rules of grammar dictate acceptable and meaningful arrangements of parts of speech. If you've got the arrangement of words right and make a sentence, it will often (but not always) still be a sentence when you substitute another word that is the same part of speech.

We told you to arrange your word strips in one color sequence, but you can figure out other sequences that make sentences. You might start by playing with very short sequences that make sentences. For instance, "noun/verb" (yellow strip/red strip) may make a simple sentence. How can you add words to that sequence and have it remain a sentence?

When you do this, you're experimenting with linguistics. One thing that linguists examine is *syntax*, the arrangement of words to produce meaning. Why is it that one arrangement of words communicates an idea, but another one doesn't? Working out the formal grammar for a language includes figuring out what makes an acceptable sentence. To do this, linguists try to build from the simplest possible sentence—just as you can do with your word strips.

Watt Due Ewe Here?

Some sentences don't mean a thing—until you read them out loud.

What Do I Need?

- The poem on this page

What Do I Do?

1 Look at the poem below. Do the words make any sense?

May Rehab All It All Am

May rehab all it all am,
Itch fleas wore widest know.
An ever wart at marry wend,
Doll am us shore two goal.

Hit fall odor twos cool wand hey,
Witch wash again stair ooze.
Hit maid a chill drench Lapland pay,
Doozy alarm hat's cool.

Wow! I Didn't Know That

When writer Sylvia Wright was young, she heard a beautiful Scottish ballad. Her favorite part was: "They have slain the Earl of Murray, and Lady Mondegreen." Ms. Wright though it was sad that Lady Mondegreen had died. Years later, she found out that the line actually was: "They have slain the Earl of Murray, and laid him on the green." In 1954, Ms. Wright wrote about her mistake. Ever since then, mishearing words has been called a "mondegreen."

2 Ask a grown-up to read the poem out loud. Don't think about the meaning of the words you see. Instead, listen to the sounds. They'll combine to make a familiar nursery rhyme.

What's Going On?

Do you know this nursery rhyme? (Here's a hint: it's about a little girl and her pet lamb.)

In this version of "Mary Had a Little Lamb," we put common English words together so that they sound like other words when you read them aloud. We got this idea from a man named Howard L. Chace. Back in the 1950s, Chace wrote "Ladle Rat Rotten Hut," the story of a little girl in a red cloak who is menaced by a "wicket woof" on her way to the house of her "groin murder."

Why play games with words and sounds?

All this relates to the human tendency to try to make sense of the world. When you hear someone talking, you do your best to find meaning in the sounds. People are always looking for (and finding) familiar forms in random shapes. That's why we can see pictures in inkblots and a face on the moon's surface.

When we search for order and meaning in sounds, the meaning we find isn't always the meaning that was intended. Children all over America salute the flag each morning with a jumble of words. "I led the pigeons to the flag," some mutter, while others salute "the republic for Richard Stans."

What Do You Call It?

Sometimes, the word you use for something is completely different from the word another person uses for the same thing—even though you speak the same language. In England, the hood of a car is called the "bonnet," French fries are called "chips," and potato chips are called "crisps."

But you don't have to go as far as England to find people who use different words for things. You can probably do that right in your own neighborhood. What do you call the biggest piece of furniture in your living room—the one that three people can sit on at the same time?

Around the Exploratorium, most people call it a "couch"—but some call it a "sofa."

The word that someone uses for something can often give you a clue about where that person lives, or where he or she grew up. People who grew up on the East Coast may use different words than people who grew up in the South or in the West. People who live in a city and people who live in the country may have different names for the same thing.

On this Field Trip, instead of taking a walk and looking for things, you get to ask the people around you to share the words that they use for ordinary objects. Show the pictures on these pages to your friends and neighbors, and ask them:

1. What do you call this?
2. Where and when did you grow up?

Write down the answers you get in a notebook or on a big piece of paper. To get a lot of different answers, you'll need to ask people of different ages who grew up in different places.

When you buy a can of a carbonated drink (like a cola or a root beer), what do you call it? Do your neighbors call it the same thing? How about your teacher?

If you're really hungry, you might have a big sandwich on a roll filled with a lot of different meats and cheeses. A lot of people call it a 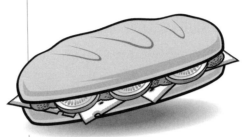 "submarine sandwich" or a "sub." How many other names for this same sandwich can you discover?

When you go to the beach or to the pool, you may wear these rubber sandals on your feet. What do you call them? Do other people at the pool call them different names?

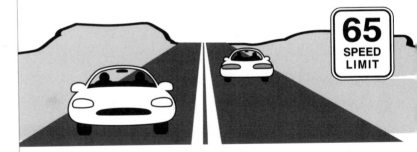

Does your school's playground have one of these? What do you call it? What do teachers call it? Do your parents have a different name for it?

You may have these for breakfast, with syrup and butter. What do you call them? How many other names for this same tasty food can you come up with?

If you go on a long trip (or maybe just to the mall), you'll probably drive on a big road with overpasses and exits, but no cross-streets. At the Exploratorium, most of us call it a "freeway." What do you call it? If you travel a long way on your trip, do people in other states call the road by other names?

For lunch, a lot of kids eat one of these thin-skinned sausages on a bun with mustard and relish. What do you call it? Do your friends call it the same thing?

The Science-at-Home Team Talks About Words

After the Science-at-Home Team tried this experiment on our own friends, we went out to lunch and talked about what we discovered: lots of ordinary things have more than one name. We also found out that you can sometimes tell how old a person is by the kinds of words they use. Linda said her mother called the big piece of living room furniture a "divan," and her grandfather called it the "davenport." Pat added that her grandfather called a suitcase a "valise."

Ellen pointed out that an older word can get passed down as a kind of family tradition, too. When her grandmother was growing up, the family kept their food in a wooden cabinet that had a big block of ice in the top to keep the food cold. They called it the "icebox." When Ellen's mother was a kid, their kitchen had an electric refrigerator, but the family kept on calling it the "icebox." Ellen says she calls her cold-food-keeper the "fridge" now, but her sister still calls it the "icebox."

We were in the middle of talking about couches and iceboxes and sub sandwiches when the waitress came. Ellen ordered a milkshake, and Linda laughed and said, "There's another one!" She told us that in Boston, where she went to school, a milkshake is just milk with flavored syrup that is shaken. If you want a drink made with milk and ice cream, you have to ask for a "frappe."

If you want to continue your own Field Trip, here's a list of some other common things that most people have, but may call by different names. You can draw your own pictures, or cut pictures out of magazines. Pick the ones that interest you—or think of your own—and ask your friends and neighbors what they call them. You'll be surprised how many simple, ordinary things have lots of different names!

- donut
- suitcase
- tights
- subway
- jungle gym
- basement
- chocolate sprinkles
- rubber boots
- frying pan
- refrigerator
- garbage

Sink or Swim

The first time you went swimming, you probably learned how to float. Floating on top of the water is one of the first steps in learning to swim. But Mary Miller, a science writer at the Exploratorium who does research in marine biology, is a scuba diver. One of the first things she had to learn was how to sink.

"People are naturally buoyant—they float easily because of the fat under their skin and the air in their lungs," she says. "When I put on a wet suit—a layer of air sandwiched between two pieces of rubber—I float even better. But when I'm scuba diving, I want to go under the water, not stay on top of it."

The air tanks that Mary and other divers wear are heavy, but not heavy enough to keep her from floating. To do that, she wears a weight belt filled with twenty-five pounds of lead. "The goal is to have enough extra weight when you're in the water so you don't float or sink," she says. "You want to be neutral, so you can stay in one place, underwater, without going up and down. It's like being weightless."

Sometimes Mary dives just under the surface of the water—about 10 feet down—to observe sea anemones, lobsters, starfish, and other marine life. But sometimes she wants to watch how fish or sharks live, and for that she needs to dive as far down as 60 feet.

At that level, the pressure of the water compresses the air in her wet suit and makes her less buoyant—she tends to sink. If she loses too much buoyancy, it will be hard to swim back up to the surface. So Mary also wears an air vest. When she blows into a tube on the vest, it inflates with air. The air in the vest makes her more buoyant. As she swims closer to the surface, she can let the air out of the vest again.

In this chapter, you won't learn how to scuba dive, but you will learn how you can use your bathtub to experiment with the way things sink and float. You'll also discover how to make a kind of "Lava Lite" out of salt and oil, how to make a sunken egg rise without touching it, and how to make a beautiful globe full of liquid swirls!

Fun in the Tub

Here are three experiments you can do in your bathtub.

What Do I Need?

- A bathtub full of water
- An unopened can of root beer or other soda
- An unopened can of the same brand of diet soda
- An orange
- Modeling clay/plasticine (don't use Play-Doh)
- A small toy or other small object

Tips for Home Scientists

You can do these experiments in water of any temperature and they won't mess up the bathtub. So you can fill the tub with warm water, do your experiments, then take a bath. If you don't have a bathtub, you can do these experiments in a bucket or a deep sink. You'll need water that's 9 to 10 inches deep.

What Do I Do?

Root Beer Float

1 Put a can of root beer (or other soda) into the bathtub. Lay it on its side on the bottom of the tub. Let go. What happens?

2 Now put a can of diet root beer (or other diet soda) into the tub. Lay it on its side on the bottom of the tub. Let go. What happens?

Wow! I Didn't Know That!

Ivory Soap—the famous floating soap—was created more than 100 years ago by a worker who accidentally left a soap-mixing machine on too long. When he made the bubble-filled mess into bars of soap, there was so much air in them that they floated. Everyone loved it! Why? In the 1870s a lot of people still bathed in rivers. Floating soap didn't sink or get lost in muddy water.

What's Going On?

Why does a can of diet soda float? Why does a can of regular soda sink?

When you put a can of soda in the water, it pushes aside, or *displaces*, a certain volume of water. If the can of soda weighs more than the water it displaces, then the can sinks. If the can of soda weighs less than the water it displaces, then the can floats.

Regular soda is heavier than the same volume of diet soda because there's lots of sugar dissolved in the regular soda. Cans of both diet soda and regular soda tend to stand up because there's air trapped in the can. The air tries to rise through the water, but it can only get as far as the top of the can. This trapped pocket of air keeps the can upright.

Another way to describe why something floats is to talk about its *density*. Density is a way of measuring how compact something is. If you divide how much an object weighs by how much space the object takes up, you get the object's density, which can be measured in pounds per quart or grams per cubic centimeter or any other weight divided by a volume.

The density of water at 4 degrees Celsius (39 degrees Fahrenheit) is 1 gram per cubic centimeter, or about 1 pound per pint. Things that are more dense than liquid water will sink in water. Things that are less dense will float. That's why ice floats. When water crystallizes into ice, it expands. Since the weight hasn't changed, the ice is less dense than liquid water.

Density can be a tough concept to understand. It isn't just a measurement of how light something is; it's a measurement of weight and volume together. A toothpick and a pine tree may have the same density. The tree is a lot heavier than the toothpick, but it also takes up more space. Both the toothpick and the tree are less dense than water, so both will float.

The Gould-Tieme family thought it was interesting to experiment with floating foods. Besides oranges, they tried floating grapes, bananas, apples, pears, carrots, and potatoes. In the McClatchey family, three-year-old Forester's favorite part of this activity was getting to eat oranges in the bathtub.

Bobbing for Oranges

1 Put a whole, unpeeled orange into the bathtub. Does it sink or float? Can you push it under the water? What happens when you let go?

2 Take the orange out of the tub. Use your fingers to peel off the rind and any thick white layer underneath.

3 Put the peeled orange back into the tub. Does it sink or float? Now put a piece of orange peel in the water. Does it sink or float?

What's Going On?

Why does an unpeeled orange float?

The orange peel is much less dense than water—it takes up a lot of space for what it weighs. So the orange peel acts like a life jacket, keeping the orange afloat.

When you peeled your orange, did it sink or float?

A dry, pulpy orange may float, even without its peel. A sweet, juicy orange will probably sink. If your orange is denser than water, it will sink; if it's less dense than water, it will float.

Winemakers use the buoyancy of grapes as a way to measure their sugar content. Sweeter grapes are denser (just like the regular soda), and they tend to sink.

Home Scientists in the Wallace family built boats of many different shapes and sizes. Then they filled the boats to see how much weight they could hold before sinking. The Stahl family made boats out of paper! The paper boats floated really well—until they got too wet. Then they sank.

Fleet of Clay

1 Form some modeling clay into a lump about the size of a golf ball. Put it in the bathtub. Does it sink or float? If it sinks, what could you do to make it float?

2 Take the lump of clay out of the tub. Flatten it in your hands to make a pancake. Pinch around the edges to make a shallow boat.

3 Carefully put the clay boat into the water. Will it float?

4 Push down on the boat with your finger. How hard do you have to push before the boat sinks?

5 Can you make a clay boat that will hold a rubber animal or a small rock and still float?

What's Going On?

Why does the clay float when I shape it into a boat?

When you shape the clay into a boat, you make a shape that displaces more water than a ball of clay. As long as your boat weighs less than the water it pushes aside, the boat floats. Some of the water is being displaced by the air that is inside the boat and under the waterline. The clay acts as a barrier that keeps the water from flowing into that air-filled space in the middle of the boat.

Salt Volcano

Make your own miniature "Lava Lite."

What Do I Need?

- A glass jar or clear drinking glass
- Vegetable oil
- Salt
- Water
- Food coloring (if you want)

DANGER!
Don't forget to be careful with glass.

What Do I Do?

1 Pour about 3 inches of water into the jar.

2 Pour about 1/3 cup of vegetable oil into the jar. When everything settles, is the oil on top of the water or underneath it?

3 If you want, add one drop of food coloring to the jar. What happens? Is the drop in the oil or in the water? Does the color spread?

Wow! I Didn't Know That!

Lava Lites are lamps that were invented by an English man named Craven Walker in 1964. They are basically tall thin glass jars filled with liquid and a special kind of colored wax, set on top of a base with a light bulb. When the bulb is turned on, the lamp glows, the liquid heats up, and the wax begins to melt. Blobs of wax rise to the top of the lamp, then cool and sink back down—over and over again.

4 Shake salt on top of the oil while you count slowly to 5. Wow! What happens to the food coloring? What happens to the salt?

Six-year-old Nina Gumkowsky shared this activity with the other students in her first-grade class. Everyone loved it! They did it over and over again and kept trying to touch the layers. It was messy, but it was fun!

5 Add more salt to keep the action going for as long as you want.

What's Going On?

Why does the oil float on the water?

Oil floats on water because a drop of oil is lighter than a drop of water the same size. Another way of saying this is to say that water is denser than oil. Density is a measurement of how much a given volume of something weighs. Things that are less dense than water will float in water. Things that are more dense than water will sink.

Even though oil and water are both liquids, they are what chemists call *immiscible liquids*. That's a fancy word that means they don't mix.

What happens when I pour salt on the oil?

Salt is heavier than water, so when you pour salt on the oil, it sinks to the bottom of the mixture, carrying a blob of oil with it. In the water, the salt starts to dissolve. As it dissolves, the salt releases the oil, which floats back up to the top of the water.

This looks like a Lava Lite. How does a Lava Lite work?

Like your oil and water, the "lava" in a Lava Lite doesn't mix with the liquid that surrounds it. When it's cool, the "lava" is a little bit denser than the liquid surrounding it. When the "lava" rests on the bottom of the Lava Lite, the light bulb in the lamp warms it

up. As it warms up, the "lava" expands a little. When it expands, the "lava" stays the same weight but it takes up more space—so it's less dense. When it's warm enough, the "lava" is less dense than the surrounding liquid, and so it rises up to the top to float. At the top of the lamp, it cools down, becomes more dense, and sinks once again. This cycle repeats over and over as the "lava" warms up and rises, then cools down and sinks.

Where did this experiment come from, anyway?

Exploratorium Teacher-in-Residence Eric Muller created this activity while playing with his food in a Chinese restaurant.

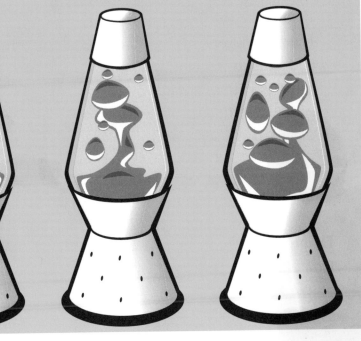

Glitter Globe

Make a fantastic toy that shimmers when you shake it.

What Do I Need?

- Rubbing (isopropyl) alcohol
- Vegetable oil
- A plastic container or glass jar with an interesting shape (long, skinny olive jars and the fancy jars that hold some marmalades, jams, or jellies work well)
- Small beads, sequins, glitter, or other tiny, shiny things
- Food coloring (if you want)
- Clear tape (if you want)

DANGER!
Don't forget to be careful with glass.

What Do I Do?

1 Fill about 1/4 of the jar with rubbing alcohol. Add a drop of food coloring,

2 Pour vegetable oil into the jar. Leave about 1/2 an inch of air at the top of the jar. Let the globs of oil settle. Is the oil on top of the alcohol or underneath it?

3 Drop tiny, shiny things into the jar. Use as many as you want. Don't use anything too heavy—like a marble—that might break the jar when you shake it.

4 When all the tiny things are in the jar, carefully pour in more oil until the jar is completely full—right up to the rim.

5 Screw the lid of the jar on very tightly. (If you want, you can tape around the lid to make sure it won't leak.)

6 Gently shake the jar. The oil and alcohol will mix and turn a milky color, and the beads and glitter will float and spin.

7 Let the oil settle again. That will take about 5 or 10 minutes. Now spin the jar instead of shaking it. What happens?

What's Going On?

Why doesn't the oil float on top of the alcohol?

Since oil floats on top of water, you might have thought that oil would float on top of alcohol, too. But the oil sinks to the bottom and the alcohol floats on top of the oil. Even though water and alcohol are both clear liquids, they have different densities. Alcohol floats on top of oil because a drop of alcohol is lighter than a drop of oil the same size.

Why don't oil and alcohol mix? For that matter, why don't oil and water mix?

The answers to these questions have to do with the molecules that make up oil, water, and alcohol. Molecules are made up of atoms, and atoms are made up of positively charged protons, negatively charged electrons, and uncharged neutrons.

The atoms that make up water molecules and alcohol molecules are arranged so that there is more positive charge in one part of the molecule and more negative charge in another part of the molecule. Molecules like this are called *polar molecules.*

The charged particles in an oil molecule are distributed more or less evenly throughout the molecule. Molecules like this are called *nonpolar molecules.*

Polar molecules like to stick together. That's because positive charges attract negative charges. So the positive part of a polar molecule attracts the negative part of another polar molecule, and the two molecules tend to stay together. When you try to mix water and oil or alcohol and oil, the polar molecules stick together, keeping the oil molecules from getting between them—and the two don't mix. When you try to mix water and alcohol, they mix fine, since both are made of polar molecules.

What's this pretty toy doing in a set of science experiments? It seems more like an art project to me.

When you make a Glitter Globe, you're experimenting with two liquids that won't mix with each other—alcohol and oil. Playing with the Glitter Globe gives you a chance to watch how liquids flow. And in the process, you make something that's pretty.

Some people think that science and art have very little in common. At the Exploratorium, we disagree. Both artists and scientists start their work by noticing something interesting or unusual in the world around them. Both artists and scientists experiment with the things they have noticed. Art and science begin in the same place—with noticing and experimenting.

Polar water molecules electrically pull together, keeping nonpolar oil molecules from getting in between.

The Amazing Water Trick

Do hot water and cold water mix?

What Do I Need?

- Two identical small, wide-mouthed jars (baby food jars are perfect)
- Hot water
- Cold water
- Food coloring
- Index cards or squares of waxed paper
- Scissors
- A large, shallow baking pan (if you don't have one, do this activity in the sink—it can be messy)

Tips for Home Scientists
This experiment can be tricky—and messy. You may want to practice step 6. Get a grown-up to help.

What Do I Do?

1 Fill one of the jars with very hot tap water. Add a drop of red food coloring. What happens to the drop? Watch for a minute, then put the red jar into the baking pan.

2 Fill the other jar with cold water. Add a drop of blue food coloring. What happens to that drop?

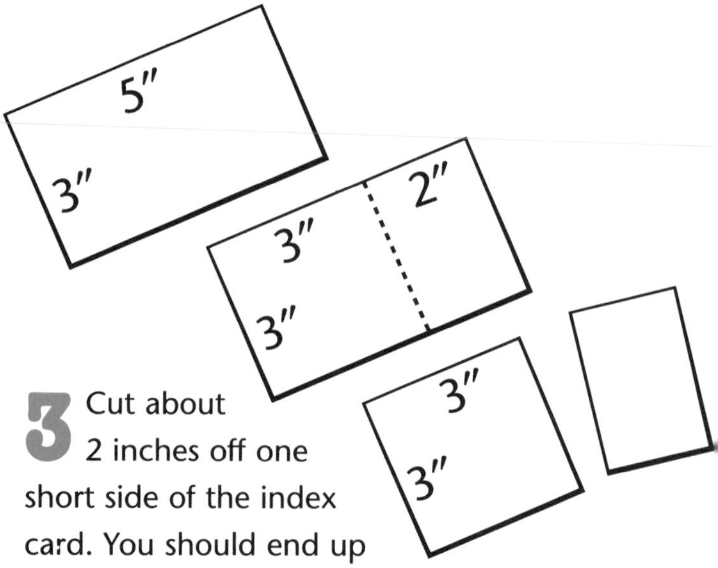

5"

3"

3"

3"

2"

3"

3"

3 Cut about 2 inches off one short side of the index card. You should end up with a square about 3 inches on a side.

4 Slowly add more water to the blue jar until you can see a bulge of water over the rim of the jar.

5 Lay the square card carefully onto the top of the blue jar. Tap the card gently with your finger. (Don't poke it. You want the card to be flat and form a seal with the water and the jar.)

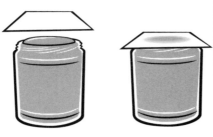

6 This part is very tricky. You may want to practice it a few times over the sink with a jar of plain water.

Pick up the blue jar and turn it straight upside-down. You don't need to put your hand on the card. The water will hold the card in place. (Just flip the jar over. Don't hesitate. If the jar is tilted but not turned over completely, the water will gush out and make a mess.)

7 Put the upside-down blue jar right on top of the red jar.

8 Have someone hold onto both jars while you very slowly and carefully pull the card out.

9 What happens? What color is the water in the top jar? What color is the water in the bottom jar?

10 Empty both jars. Rinse them. Repeat steps 1 through 6—but put the jar with the blue-colored cold water in the baking pan and put the card on top of the jar with the red-colored hot water. Turn the red jar upside-down and put it on top of the blue jar.

11 Slowly pull out the index card. What happens? What color is the water in the top jar? What color is the water in the bottom jar?

What's Going On?

What if I don't have any baby food jars?

If you don't have any baby food jars and you don't have a baby who will empty some jars for you, we suggest that you buy a jar of strained peaches or applesauce baby food. You can empty the jars over vanilla ice cream for a tasty dessert or mix them in a blender with milk and ice to make a smoothie.

If you don't want to buy baby food, we have also done this experiment using two identical small, clear, drinking glasses. You just have to make sure that the mouths of the glasses match up perfectly, without a leak.

Why does the water mix so quickly when the glass of hot water is on the bottom?

If you've already made a Salt Volcano (page 104) or a Glitter Globe (page 106), you probably know that some liquids float on top of other liquids. Oil floats on water. Alcohol floats on oil. That's because these liquids have different densities. Whenever you put together two liquids that don't mix, the liquid that is less dense will float on top of the denser liquid. A drop of oil weighs less than a drop of water the same size. The oil is less dense than the water, so it rises to the top.

When you heat up water, the water molecules start moving around faster and faster. They bounce off each other and move farther apart. Because there's more space between the molecules, a volume of hot water has fewer molecules in it and weighs a little bit less than the same volume of cold water. So hot water is less dense than cold water. When you put the two together with the hot water on the bottom, the hot water rises to the top, mixing with the cold water along the way and creating purple water.

Why doesn't the water mix when the hot water is on top?

When the cold water is on the bottom, the hot water doesn't have to rise—it's already on top. The cold blue water stays on the bottom and the hot red water stays on top.

How can I experiment further?

If you try Submarine Egg (page 111), you'll find out that salt water is denser than water without salt. What do you think would happen if you tried this experiment

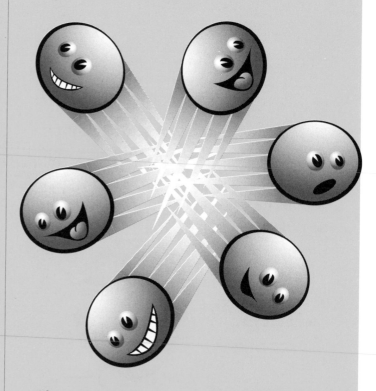

with a jar of salt water on top and a jar of water without salt on the bottom? Try it and see. Use food coloring to color the salt water a different color than the plain water so that you can see what happens.

Submarine Egg

Make an egg rise through the water—without touching it!

What Do I Need?

- A raw egg
- A drinking glass, wide-mouth jar, or deep glass bowl
- Water
- Salt

DANGER!
Don't forget to be careful with glass.

What Do I Do?

1 Fill the glass or jar about 2/3 full of water.

2 Very carefully put the egg into the jar. Does the egg float in the water or sink to the bottom?

3 Pour about 1/4 cup of salt into the jar and gently stir to dissolve the salt. What happens to the egg? (If nothing happens, add another 1/4 cup of salt.)

Wow! I Didn't Know That!

Between England and Denmark is a body of water called the North Sea. Lots of rivers flow into the North Sea, so in some places there's a layer of fresh water that sits on top of the heavier salt water. Codfish eggs—which are lighter than salt water but heavier than fresh water—float at the boundary in between.

What's Going On?

Why does an egg sink in fresh water and float in salt water?

When you put an egg in the water, it displaces a certain volume of water. Since the egg weighs more than the water it displaces, the egg sinks. When you add salt to the water, you don't change the egg. But you do add to the water's density. A quart of salt water weighs more than a quart of fresh water—it's more dense. The egg is lighter than the salt water it displaces—so it floats.

Go With the Flow

Spin the bottle to see beautiful swirling shapes.

What Do I Need?

- A clear plastic bottle or jar with a tight-fitting, screw-on cap or lid (a clear plastic water bottle works great)
- Liquid hand soap that has glycol stearate in it (The brand we used is Colgate-Palmolive's Softsoap, but any brand of liquid soap with glycol stearate—not glycol distearate— will work; check the ingredients on the label.)
- Water
- Food coloring
- Clear tape

What Do I Do?

1 Fill the bottle or jar about 1/4 full with liquid soap. Add a drop or two of food coloring. The coloring will make the swirls easier to see.

2 Turn on your faucet so you have just a trickle of water. Use that to fill up the rest of the bottle. (If you run the water too hard, you'll get foam.) Make sure that the water fills the bottle all the way to the very top.

3 Screw the cap on the bottle. Turn the bottle upside-down a few times to mix the soap and water. If you get foam, take the cap off and trickle some more water into the bottle. The foam will run over the edge. Recap the bottle tightly.

4 Dry the bottle and the cap, then wrap clear tape around it so the bottle won't leak.

5 Twirl the bottle slowly. What do you see? What happens when you stop twirling the bottle? What happens if you spin it quickly?

6 Try shaking the bottle up and down or side to side. What different patterns do you see inside the bottle?

7 If the liquid inside the bottle looks like it's all one solid color, just twirl or shake it again to make more patterns. If the cap on the bottle is sealed, Go with the Flow can last for years. (Note: If you find a really pretty plastic bottle or jar, you can give this to someone as a gift.)

What's Going On?

Why can I see patterns in the water?

Normally, you can't see how the water is moving inside a full jar of water. Water that's moving in one direction looks the same as water that's moving in another direction. But glycol stearate, the chemical that gives some liquid hand soaps a pearly look, lets you see patterns flow in water.

What kinds of patterns can I see in my jar? Who cares about these patterns?

When you turn the bottle slowly, you'll probably see smooth streaks in the water. When layers of water are moving slowly and smoothly past each other, you get this pattern, which scientists call *laminar flow*.

When you suddenly stop turning the bottle, or when you turn it very fast, you may see lots of swirls and wavy patterns. When one layer of water moves rapidly past another layer of water, it causes turbulence, which you see as swirly patterns.

When people design airplanes, cars, boats, golf balls, and other things that move through air or water, they study the patterns blowing air or flowing water makes as the object moves through it. Differences in the flow of air or water can affect how well an airplane flies, how much mileage a car gets per gallon, how fast a boat can go, or how far a golf ball will fly when you smack it with a club.

Floating Around Your Neighborhood

If you live near water—a lake, a river, or the ocean—it's easy to see a lot of things that are floating. Ducks float. So do boats. You might notice that a boat full of people or cargo floats much lower in the water than an empty boat does. If you watch a leaf that has fallen into a pond, the way it floats can reveal the patterns of the moving water.

But you don't have to live near water—you can find floating things no matter where you live. A dry sponge will float in your sink because it's less dense than the water around it. Some bars of soap have a lot of air in them and will float in your bathtub. Explore your own neighborhood and look for things that float. You'll find lots of them—in places that may surprise you!

The Science-at-Home Team Sinks and Floats

One morning, the Science-at-Home Team decided to get together before work, then walk to the Exploratorium and look for floating things. We all met at a cafe near the museum. As soon as the waiter brought our food, we realized we didn't have to wait until we were outside to watch how things float— we could watch our breakfast.

Ellen's hot chocolate came with a white, puffy marshmallow floating on top. The marshmallow floated because it was full of bubbles of air. Linda's latté came in a tall glass. We could all see that the denser liquid of the brown coffee was on the bottom and the foamy white steamed milk floated on top.

Pat's tea also came in a glass mug. She slowly added a few drops of cream to the clear brown liquid, and we all watched them swirl around in the mug. That reminded Pat of how the food coloring swirled in the cold water when we did The Amazing Water Trick, and reminded Linda of the pearly patterns in Go with the Flow. Pat added a few more drops of cream, and the swirling patterns continued until she stirred everything together with her spoon.

It was a warm morning, so Ellen ordered a glass of water with ice. When the waiter brought it to the table, we saw that the clear ice cubes were floating. Ice cubes are just frozen water. Why would they float in liquid water? Linda explained that when water freezes, its volume expands just a little bit—about 10 percent. That means ice is less dense than water, so ice cubes float.

We were still pondering this when our food came. Pat noticed that there was melting butter on top of her bowl of oatmeal. When she poured milk into the bowl, the fat in the butter floated on top, just like the oil did in Salt Volcano. We tried to get her to put lots of salt on her oatmeal, to see if the butter would sink, but she refused.

Ellen and Linda both ordered cereal. Linda had Cheerios, and Ellen had granola. Right away we saw that the Cheerios floated in milk, but the granola bits didn't. After a few minutes, though, Linda's Cheerios got soggy, and they sank to the bottom of her bowl, too.

When we were paying the bill, we saw the waiter setting a table up for lunch. He put a cruet with an oil-and-vinegar dressing on the table, and Ellen noticed that the oil was floating on top of the vinegar. Since oil and vinegar don't mix, she said, most vinaigrette dressings have an ingredient like mustard that helps form an *emulsion* when the dressing is shaken. An emulsion is created when tiny drops of one liquid are scattered throughout another liquid. The mustard helps keep the oil and vinegar from separating when the dressing is poured over lettuce. Ellen promised to give us her favorite salad dressing recipe when we got back into the Exploratorium.

As we were walking back toward the museum, we saw some beautiful puffy clouds floating high up in the air. Pat explained that they were cumulus clouds, formed by rising warm air. As the air rises, it expands. The air also cools off as it expands, and the water vapor condenses into the cloud that we saw.

Pat said that's also the way hot air balloons work. The air inside the balloon is heated, so it expands. The balloon floats through the sky because the air inside it is less dense than the air around it. She said that since helium is a gas that's lighter than air, a helium balloon will also float up into the sky if you let go of its string.

Linda agreed and said that sometimes just filling something with air will make it float in water. She said that the last time she'd taken a soak in a friend's hot tub, she took a big, deep breath and filled her lungs with air, and her body floated up in the water. When she exhaled, her body sank again. We talked for a while about how air inside something can make it float—like the air in Linda's lungs, the air inside an inner tube, or the air inside the hull of a boat.

When we got to the lagoon next to the museum, we stopped for a while to watch the ducks. They made Pat think about cormorants and grebes—seabirds that you can see on San Francisco Bay. Cormorants and grebes float on the water, but they don't float as well as ducks. A grebe looks a little like an overloaded ship, floating half-submerged in the water. Pat said that ducks feed by sticking their heads underwater, but grebes dive all the way under the water for their food. She speculated that grebes might be like scuba divers—if they floated too well, they wouldn't be able to dive under water.

Duck

Grebe

Just then, a duck rolled its head underwater. When it came up again, Ellen noticed that beads of water rolled off the duck's feathers. Linda said that ducks' feathers are oily. Since oil and water don't mix, the oil keeps the ducks' feathers from getting soggy and heavy. We were surprised at all the places around the Exploratorium that we could watch things sink and float. What can you find in your neighborhood?

Ellen's vinaigrette recipe
1½ tablespoons red wine vinegar
1 teaspoon Dijon mustard
1/4 cup plus 2 tablespoons olive oil
1/4 teaspoon salt
dash of pepper

Combine vinegar, mustard, salt, and pepper in a bowl. Whisk in oil until blended. Serve over your favorite salad ingredients. Makes 1/2 cup.

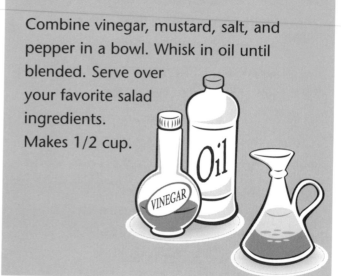

Exploring Further

Tips for Teachers and Suggested Reading

Tips for Teachers

Although the *Science Explorer Out and About* was designed for families, the same qualities that make these activities easy to do at home also make them perfect for the classroom: the ingredients are simple to use, easy to find, and inexpensive to buy. Every experiment in this book is not only fun to do, but it's also based on solid science, or a combination of art and science. The information provided in each section can help support kids as they learn (without even noticing!) about chemical bonding, the nature of light, architectural engineering, linguistics, the principles of density and fluid mechanics, and much, much more.

At the Exploratorium, we've found that kids learn best when they're actively participating in the discovery process. All of the activities in the *Science Explorer* series foster skills in creativity, problem solving, and critical thinking—especially when kids are given the opportunity to explore on their own. We've also found that these activities are great for helping kids learn to work cooperatively, as each child can use his or her strengths to support the others in the group. Even children whose primary language is not English can be successful participating in these activities.

Suggested Reading

The Science Explorer by Pat Murphy, Ellen Klages, Linda Shore, and the staff of the Exploratorium (New York: Henry Holt and Company, 1996). Don't miss the first book in this exciting, family-focused, science activity series!

If you'd like to keep on exploring, here are some books that can help you.

Books for Kids Ages 7 to 9

1. It's All Done with Mirrors
Science with Light and Mirrors by Kate Woodward (London: Usborne Publishing Ltd., 1991). Science activities that let kids explore properties of light. Includes explanatory notes for parents and teachers.

2. Surprising Structures
What It Feels Like to Be a Building by Forrest Wilson (Washington, D.C.: The Preservation Press, 1988). Using playful drawings of humans and animals and humorous text, this book shows that architecture and people have a lot in common.

3. Taking Things Apart
Everyday Machines: Amazing Devices We Take for Granted by John Kelley (Atlanta: Turner Publishing, 1995). An illustrated look at what goes on inside your hair dryer, toaster, smoke alarm, digital watch, and many other common household objects.

The Way Things Work by David Macaulay (Boston: Houghton Mifflin Co., 1988). Amazingly detailed illustrations of what's inside everyday objects—from ballpoint pens to car engines—and how they function.

4. Making Changes
Crystals and Gems by R. F. Symes and R. R. Harding (New York: Knopf, 1991). A fascinating book full of great color photos and lots of interesting information.

The Science Chef: 100 Fun Food Experiments and Recipes for Kids by John D'Amico and Karen Eich Drummond (New York: John Wiley and Sons, 1995). This book contains lots of information about the amazing changes that take place in the kitchen, and tells you how to make your own bread, cheese, and yogurt.

5. Made in the Shade
Shadows: Here, There, and Everywhere by Ron and Nancy Goor (New York: Crowell, 1981). A good beginner's book about shadows, how they're formed, and how they can reveal the shapes and textures of objects.

6. What Do You Say?
Word Works: Why the Alphabet is a Kid's Best Friend by Cathryn Berger Kaye (a Brown Paper School Book; Boston: Little, Brown and Company, 1985). A great book that covers everything linguistic—from word games and grammar to signs, symbols, and poetry.

7. Sink or Swim
Experiment with Water by Bryan Murphy (Florida: Action Publishing, 1991). A fun, heavily illustrated book for kids that's full of information about and experiments with water.

Books for Older Kids and Grown-ups without a Science Background

It's All Done with Mirrors by Irvin D. Gluck (New York: Doubleday & Company, Inc., 1968). For anyone who wonders how mirrors work and what they can do—a book of explanations, experiments, and projects.

Buckminster Fuller by Robert R. Potter (New Jersey: Silver Burdett Press, 1990). This biography of geodesic dome pioneer Buckminster Fuller traces Fuller's life and work and the many obstacles he faced to reach his goals.

The Book of Buildings by Richard Reid (Chicago: Rand, McNally & Company, 1980). A comprehensive, heavily illustrated guide to the architecture of Europe and North America from early classical times to the present.

Janice VanCleave's Machines by Janice VanCleave (New York: John Wiley & Sons, Inc., 1993). Twenty simple and fun experiments, plus dozens of suggestions on how to develop your own science fair project.

The Usborne Illustrated Dictionary Series (London: Usborne Publishing, Ltd.): *The Dictionary of Biology,* by C. Stockley (1986); *Dictionary of Physics,* by C. Oxlade, C. Stockley, and J. Wertheim (1986); and *Dictionary of Chemistry,* by J. Wertheim, C. Oxlade, J. Waterhouse (1986). Beautifully illustrated and clearly written, these books provide explanations of basic science concepts.

Sundials: Their Construction and Use by R. Newton Mayall and Margaret W. Mayall, third ed. (Cambridge: Sky Publishing Corp., 1994). Older kids can learn how a sundial works and get instructions for designing and constructing several different permanent and portable dials.

The Oxford Guide to Word Games by Tony Augarde (New York: Oxford University Press, 1984). Ways to play with words, letters, and language fill this interesting compendium.

Advanced Books for Grown-ups with a Science Background

Towards a New Architecture by Le Corbusier, translated from the 13th French edition with an introduction by Frederick Etchells (New York: Dover Publications, Inc., 1986). One of this century's master builders, Le Corbusier discusses his goal of achieving a pure creation of the spirit.

Basic Machines and How They Work prepared by the Bureau of Naval Personnel (New York: Dover Publications, Inc., 1971). Covers basic theory ranging from the lever and inclined plane up through basic computer mechanisms. This book requires no more than an elementary knowledge of mathematics.

The Language Instinct: How the Mind Creates Language by Steven Pinker, (New York: William Morrow and Company, 1994). Everything you always wanted to know about language: how it works, how children learn it, how it changes, how the brain computes it, and how it evolved.

Many of these books are available from the ExploraStore, where you can shop in person or order by phone at 415-561-0393. For online information about Exploratorium publications and products, see our Web site at http://www.exploratorium.edu.

About the Exploratorium

and the Science-at-Home Project

What Is the Exploratorium?

In a residential neighborhood near the Golden Gate Bridge there's a building filled with flashing lights, machines that buzz and whir, and a constant hum of excited conversation. It's the Exploratorium, San Francisco's world-renowned museum of science, art, and human perception.

Physicist Frank Oppenheimer opened the museum in 1969 as a place that would introduce people to science by letting them explore and ask questions. The mission of his new museum was to let people discover that science is not only understandable, but exciting and fun.

Today, the Exploratorium has more than 600 exhibits—and they all run on curiosity. You don't just look at these exhibits, you play with them. You can touch a tornado, build a bridge, look inside a cow's eye, or leave your shadow on a wall. Every year, more than half a million people from all over the world—including thousands of kids on school field trips—come to the Exploratorium to discover for themselves how the world works.

What Is the Science-at-Home Project?

This book is the second in a series of Exploratorium Science-at-Home books. Over the years, the teachers and scientists and writers who work at the Exploratorium have created classroom-sized versions of many of the museum's exhibits. Like *The Science Explorer,* the first book in this series, many of the experiments in this book are "home versions" of Exploratorium exhibits, designed for families to do using simple materials found at home or in any grocery store.

The Science Explorer Out and About is full of activities that let you make things, take things apart, experiment, and see what happens—in other words, play around. At the Exploratorium, we think that the best way to learn about science is to have fun.

Each of the experiments in this book was developed at the Exploratorium, then tested by families all over the world. They reported back to us, telling us which experiments worked for them—and which ones didn't—sharing their own experiences and making suggestions about changes or improvements they'd like to see. We used their comments and suggestions to put this book together.

The Science-at-Home project hasn't ended with the publication of this book. We've already started work on the third book in this series! If you would like to let us know what you thought of this book, you can contact us at the address on page xii.

Visit the Exploratorium On-Line at http://www.exploratorium.edu

You can find more family fun from the Exploratorium at our Web site: http://www.exploratorium.edu. Once inside the "virtual Exploratorium," you can find out more about the museum and its historic home in the Palace of Fine Arts in San Francisco or discover what's hot in the ExploraStore. You can play with online exhibits, sample our publications, find classroom activities, or link to other interesting places on the Internet.

Or, Even Better, Come and Visit Us!

The Exploratorium is located in the landmark Palace of Fine Arts in the Marina District of San Francisco, just south of the Golden Gate Bridge. The next time you're in San Francisco, come by and discover all the amazing activities, exhibits, and special events at the Exploratorium.

Thank You Very Much!

Our thanks to the hundreds of families who helped us test the activities in this book. Their spirit of creativity and discovery made a real difference to the success of this project. Without their support and encouragement, this book would not have been possible.

The Abbitt Family
The Adams Family
The Advena Family
The Anderson Family
The Asher Family
The Baggett Family
The Bahannan Family
The Bardo Family
The Barnes Family
The Barry Family
The Barton Family
The Bath Family
The Beckman Family
Bigfork Elementary School, Grade 4, Section K
The Blair-Stolk Family
The Blumenfield Family
The Brawner Family
The Brenning Family
The Brewer Family
The Briggs Family
The Brock-Knight Family
The Bulaevsky Family
The Buoni Family
The Burns Family
The Burrell Family
The Campbell Family
The Cantrell-Dresser Family
The Carlson Family
The Carpenter Family
The Chase Family
The Ching Family
The Cifor Family
The Cohen Family
The Cote Family
The Dahl Family
The Dean Family
The Dorland-Junier Family
The Dungy-Maclean Family
The Durham-Williams Family
The Eiles Family

The Elahmadie Family
The Ellerton Family
The Elliser Family
The Fairchild Family
The Faneuf Family
The Feld Family
The Feldman Family
The Fetteroll Family
The Fischman-Quant Family
The Fisher Family
The Fisher-Hegland Family
The Flores Family
The Foote-Mandel Family
The Ford Family
The David Frost Family
The Mrs. Siobhan Frost Family
The Gertsch Family
The Giancaspro Family
The Gianfredi Family
Girl Scout Troop 2108
The Goers Family
The Golden Family
The Gould Family
The Gould-Thieme Family
The Grasso-Underwood Family
The Greenberg Family
The Greene Family
The Gumkowski Family
The Haddan Family
The Hagen Family
The Hall Family
The Hamilton Family
The Hansen-Hamilton Family
The Don Hansen Family
The J & K Hansen Family
The Harkins Family
The Harpster Family
The Hathaway Family
The Haupt Family
The Hinze Family
The Hitchner Family
The Ho Family

The Hollenbeck Family
The Jackson Family
The Johnson Family
The Johnston Family
The Judd-Clear Family
The Jue Family
The Kaborycha-Mount Family
The Kahler Family
The Kann Family
The Kielman Family
The Kim Family
The Kincaid Family
The Klinedinst Family
The Kloess Family
The Kowalczewski Family
The Kramb Family
The Kremsdorf Family
The Kroo Family
The Lamb-Chalmers Family
The Landis Family
The Lanning-Tornquist Family
The Lapid Family
The Larson Family
The Leach Family
The Lee Family
The Levy Family
The Danny Lindsey Family
The Lyons and Ching Family
The MacCormack Family
The Maguire-Sawicky Family
The Mandel Family
The Marriner-Rothschild Family
The Marshall Family
The Matherly Family
The McClatchey Family
The McGovern-Cox Family
The McKinnon-Tam Family
The Messier Family
The Emily Morris Family
The Jamie Morris Family
The Morrow Family
The Moseley Family
The Mudd Family
The Narigon Family
The Ned Cox Family
The Newton Family
The Nilsson Family
The Nunamaker Family

The O'Brien Family
The Ogden Family
The Olacke Family
The Opotow-Chang Family
The Orton Family
The Painter-Heston Family
The Palumbo Family
The Paulson Family
The Penston Family
The Pera Family
The Perry Family
The Phillips Family
The Potts Family
The Proud Family
The Provencher Family
The Prudhomme Family
The Pryden Family
The Ramsey Family
The Randall Family
The Reichman Family
The Reinwald Family
The Riefkohl Family
The Roberts Family
The Rockwood Family
The Rosano Family
The Rowen Family
The Ryan Family
The Sandoval Family
The Santi Family
The Scarla Family
The Schmitt Family
The Schreyer Family
The Schubring-Alcantara Family
The Schulte Family
The Schwartz Family
The Scornavacca-Wenn Family
The Shah Family
The Shia Family
The Sibley Family
The Smith Family
The Smith-Mancuso Family
The SooHoo Family
The Stahl Family
The Stevenson Family
The Street Family
The Struttmann Family
The Sullivan Family

The Summerhayes Family
The Taylor Family
The Theisen Family
The Thomas Family
The Thompson Family
The Trevillyan Family
The Tsang Family
The Tucek Family
The Turner Family
The Underhill Family
The Valentine Family
The Vavra Family

The Volkoff Family
The Wallace Family
The Watkins Family
The Watson Family
The Weiss Family
The Weller Family
The Widerquist Family
The Wildgrube Family
The Willisson-Rowe Family
The Woodbury Family
The Young Family

Acknowledgments

At the Exploratorium, no one works alone. The Science-at-Home Team had the help and support of many people, both inside the Exploratorium and out. The experiments and activities in this book came from years of work and play by the staff and teachers at the School in the Exploratorium, the Exploratorium Teacher Institute, and the Exploratorium's Children's Outreach Program.

Jenefer Merrill, a former member of the Exploratorium staff, a creative teacher, a writer, and a good friend to the Science-at-Home Team, helped us develop new activities and test them on her very own children, Sadie and Kate Switzer, and with family friend Nikolai Korhummelstone.

Many people at the Exploratorium helped us by telling us about their favorite activities, testing those activities, or reviewing what we wrote. We'd especially like to thank Vivian Altmann, Paul Doherty, Ken Finn, Cappy Greene, Kurt Feichtmeir, Karen Kalumuck, Barry Kluger-Bell, Fred Stein, and Eric Muller.

We also want to thank those who helped us turn a bunch of words into a book. Jason Gorski not only drew the characters, his illustrations helped us find the clearest way to present the information. Amy Snyder provided the photos, wandering the streets of San Francisco with us in search of the very best shadows. Gary Crounse, our eternally calm graphic artist, took each chapter of this book and designed a Science-at-Home packet that we could send out to our Home Scientists. Mark Nichol proofed the packets, and Megan Bury not only got them to the printer and into the mail, but also maintained our databases and analyzed the questionnaires from the participating families.

When the authors were too weary to go on, Ruth Brown came in as an editorial angel, taking the chapters we had crafted and hammering them into final text, working with David Sobel at Henry Holt and Company, who we value for both his editing acumen and his patience. Ellyn Hament made sure we caught every last typo, and Diane Burk and Stacey Luce used their graphic skills to transform the individual activity packets into this book.

The Science Explorer Out and About is a sequel to *The Science Explorer,* a book that was shaped by the advice of Phil Morrison, Phylis Morrison, and Bernie Zubrowski. Testing and development of *The Science Explorer* was made possible through the support of the Informal Science Education Program of the National Science Foundation, and the Pacific Telesis Foundation.

This book would not have been possible without the moral and administrative support of many people. Thanks to Kurt Feichtmeir for being the keeper of the budget (a thankless but necessary task), and for remaining cheerful through it all. And thanks to Rob Semper and Goéry Delacôte for giving us the institutional backing we needed.

Index

San Francisco

EXPLORaTORiUM

The Museum of Science, Art,
and Human Perception